Surveys on
Theories in
Economics and
Business Administration
Vol.
1

A Survey of Dynamic Games in Economics

T0350248

Ngo Van Long
McGill University, Canada

World Scientific

NEW JERSEY · LONDON · SINGAPORE · BEIJING · SHANGHAI · HONG KONG · TAIPEI · CHENNAI

Published by

World Scientific Publishing Co. Pte. Ltd.

5 Toh Tuck Link, Singapore 596224

USA office: 27 Warren Street, Suite 401-402, Hackensack, NJ 07601

UK office: 57 Shelton Street, Covent Garden, London WC2H 9HE

British Library Cataloguing-in-Publication Data
A catalogue record for this book is available from the British Library.

ISBN-13 978-981-4293-03-7
ISBN-10 981-4293-03-2

Typeset by Stallion Press
Email: enquiries@stallionpress.com

Printed in Singapore by World Scientific Printers.

This book is dedicated to Kim-Chau, Bach, Chi, Eric, and Bambou.

CONTENTS

PREFACE

This book originates from my years of teaching economic dynamics at McGill University and the Australian National University. I have found that most students would like to be exposed to a wide variety of dynamic game models, with applications to various fields of economics.

This book is intended to present a broad picture of dynamic games in economics. As a result, I deliberately avoid dealing with issues that would appear excessively technical for most graduate students and for economists with a general PhD training in economics. Those who prefer reading treatments of the subject at a higher level of formalism should consult graduate-level textbooks or treatises, such as Dockner *et al.* (2000), *Differential Games in Economics and Management Science*, or Başar and Olsder (1995), *Dynamic Noncooperative Game Theory*.

A decade has passed since the publication of Dockner *et al.* (2000). While the present work refers to quite a number of articles published between 2000 and 2010, it is not meant to be an update on that volume. The two volumes are complements, not substitutes. This work is also a complement to Jørgensen and Zaccour (2004), *Differential Games in Marketing*.

A good friend of mine, the late Jean-Jacques Laffont, told me a memorable story. In writing his books, *Fundamentals of Public Economics*, and *The Economics of Information and Uncertainty*, he faced an "adverse selection" problem: among many good papers and models that deal with the same subjects, which ones should he select for a detailed exposition? He decided to take the risk of appearing biased toward articles of which he was either an author or a co-author, one of the reasons being that he had a comparative advantage in explaining them. I have taken the same risk, for the same reasons. Naturally, in a survey, one must be selective: one cannot give equal treatments to equally good papers. Like a tour guide, I allocate more time (and space) to articles that I think the readers would find interesting and yet not too hard to grasp.

While the actual preparation of this book took about 18 months, it is not an exaggeration to say that its gestation period is more than three decades. During my journey, I have benefited from discussions with many friends and co-authors. They have reinforced my interest in economic modelling, dynamic optimization, and dynamic games. I would like to thank Francisco Alvarez-Cuadrado, Jean-Pierre Amigues, Venkatesh Bala, Sanjay Banerji, Joydeep Bhattacharya, Hassan Benchekroun, Richard A. Brecher, Edwin Burmeister, Jim Cassing, Carl Chiarella, Richard Cornes, Engelbert Dockner, Charles Figuières, Bruce Forster, Kenji Fujiwara, Gérard Gaudet, John M. Hartwick, Horst Herberg, Arye Hillman, Nguyen Manh Hung, Steffen Jørgensen, Larry Karp, Seiichi Katayama, Murray C. Kemp, Toru Kikuchi, Pierre Lasserre, Jean-Jacques Laffont, Didier Laussel, Daniel Léonard, Xiao Luo, Richard Manning, Stephanie McWhinnie, Michel Moreaux, Kazuo Nishimura, Koji Okuguchi, Hiroshi Ohta, John Pitchford, Horst Raff, Ray Rees, Ray Riezman, Koji Shimomura, Horst Siebert, Hans-Werner Sinn, Gerhard Sorger, Antoine Soubeyran, Raphaël Soubeyran, Frank Stähler, Hideo Suzuki, Haru Takahashi, Makoto Tawada, Mabel Tidball, Binh Tran-Nam, Stephen Turnovsky, Neil Vousden, and Kar-Yiu Wong.

Other colleagues and many of my students have provided valuable inputs, comments, and feedback. My thanks go in particular to Gustav Feichtinger, Philippe Michel, Rick van der Ploeg, Steve Salant, Henry Y. Wan, Jr., Cees Withagen, Georges Zaccour, Soham Baksi, Enrique Calfucura, Steven Chou, Engin Dalgic, Viet Do, Isabel Galiana, Octave Keutiben, Bodhi Sengupta, and Shengzu Wang.

I am particularly grateful to Murray C. Kemp, Kim-Chau Long and Kenji Fujikawa for reading the entire typescript, making corrections and offering suggestions for expositional and stylistic improvements. Needless to say, I am responsible for the remaining errors.

INTRODUCTION

Dynamic games, introduced to economics by Roos (1925, 1927), and neglected until the early 1970s, have become a standard tool of economic analysis. Indeed, one of the emerging trends in the expanding relationship between game theory and economics is the application of dynamic/stochastic games in a variety of settings. Dynamic games include differential games (set in continuous time), difference games (set in discrete time), and timing games.

While early economic articles using dynamic games mostly appeared in highly technical journals or volumes, nowadays one easily comes across papers using dynamic games in mainstream general journals such as the *American Economic Review, Journal of Political Economy, Economic Journal, Canadian Journal of Economics, European Economic Review, Oxford Economic Papers, Economica, Scandinavian Journal of Economics, Japanese Economic Review*, and in field journals such as *Journal of Public Economics, International Journal of Industrial Organization, Journal of International Economics, Journal of Public Economic Theory, Review of International Economics, Journal of Development Economics, Journal of Environmental Economics and Management, Resource and Energy Economics, Environmental Modelling and Assessment*, and many others.

The purpose of this volume is to present a survey of the development of dynamic games in economics, with special emphasis on applications. An excellent survey was provided by Clemhout and Wan (1994). Since then, there have been many economic articles with further applications of differential games. This fact alone is sufficient to justify my enterprise. I must stress that this volume is not meant to be a comprehensive treatment of differential games. Readers interested in precise definitions and fine points are referred to books on dynamic games (e.g., Başar and Olders, 1995; Dockner *et al.* 2000). The latter also contains many applications in economics and management science. I will assume that the reader has some working knowledge of optimal control theory and dynamic programming, for which there are several good books (e.g., Léonard and Long, 1992; Kamien and Schwartz, 1991).

Economists have used dynamic games to analyze a variety of problems in various fields. In this volume, I survey dynamic games in environmental economics (Chap. 2), natural resources economics (Chap. 3), international economics and development economics (Chap. 4), industrial organization (Chap. 5), public economics (Chap. 6), and macroeconomics (Chap. 7). Naturally, there are overlaps and omissions, for which I apologize.

My wish is that this book will stimulate the reader's interests in the use of dynamic games to analyze new problems or old problems from a new perspective. The next step would be to equip yourself with a pencil, a piece of paper, and a good textbook on differential game. Bon voyage!

Chapter 1

BASIC CONCEPTS

1.1. What Are Dynamic Games?

Let us begin with a preliminary characterization of dynamic games. A game involves two or more players. A player can be any decision-making entity, such as a government, a political party, a firm, a regulatory agency, an international organization, a household, or an individual. (I will use the pronoun "it" to refer to a generic player, even though in some contexts, the pronouns "he" or "she" may be more appropriate). A dynamic game extends over a time horizon (finite or infinite) and normally displays the following properties:

- the players may receive payoffs in every period (or at every point of time);
- the overall payoff for a player is the sum (or integral) of its discounted payoffs over the time horizon, possibly plus some terminal payments;
- the payoff that a player receives in a period may depend on both the actions taken in that period and the "state of the system" in that period, as represented by one or several "state variables";
- the state of the system changes over time, and the rate of change of the state variables may depend on the actions of the players, as represented by their "control variables"; and
- the rate of change of a state variable is described by a difference equation or a differential equation, often called the "transition equation" or "dynamic equation".

Thus, I will exclude from consideration "repeated games" (such as the repeated prisoners' dilemma etc.) since in such games there are no state variables and no transition equations that describe the changing environment in which the players operate.

Economists have used dynamic games to analyze a variety of problems in various fields, such as dynamic oligopoly, dynamic contributions to a public good, dynamic game of optimal tariffs and retaliation, redistributive taxation in the presence of forward-looking agents, exploitation of common property resources, non-cooperative environmental policies, and the arms race. In this survey, I intend to introduce the readers to the basic equilibrium concepts in dynamic games and present a number of interesting dynamic game models and results in various fields of economics.

This chapter introduces the basic concepts and some ideas about solution techniques. In Sec. 1.2, I introduce two main equilibrium concepts, open-loop Nash equilibrium (OLNE) and Markov-perfect Nash equilibrium (MPNE), and illustrate their difference by means of simple examples. In Sec. 1.3, I introduce the concept of hierarchical dynamic games and two equilibrium concepts for such games, the open-loop Stackelberg equilibrium (OLSE) and the feedback Stackelberg equilibrium (FBSE).

1.2. Open-loop Nash Equilibirum and Markov-perfect Nash Equilibrium

One of the most important distinctions in dynamic games is that between "open-loop strategies" (or pre-commitment strategies) and "Markov-perfect strategies" (or feedback strategies).[1] A player's open-loop strategy is a planned time path of its actions.

An OLNE is a profile of open-loop strategies (one for each player) such that each player's open-loop strategy maximizes its payoff, given the open-loop strategies of other players. Some early articles analyze dynamic games using exclusively this equilibrium concept. See, for example, Clark and Munro (1975), Salant (1976), Dasgupta and Heal (1979, Chap. 12), Spence (1979, 1981), Flaherty (1980a,b), Crawford *et al.* (1980), Reinganum (1981a, 1982a), Fudenberg and Tirole (1983), Lewis and Schmalensee (1980), Chiarella *et al.* (1984) and Loury (1986).[2] More recent articles that focus exclusively on OLNEs include Gaudet and Long (1994, 2003), Lambertini and Rossini (1998), Sorger (2002), Benchekroun *et al.* (2009, 2010), and Fujiwara and Long (2010).

---

[1] I use the terms feedback strategy and Markov-perfect strategy interchangeably. Similarly, a feedback Nash equlibrium (FBNE) and a Markov-perfect Nash equilibrium (MPNE) are alternative names for the same concept.

[2] It turns out that in the case of Reinganum (1981a, 1982a), the OLNE is also the FBNE. Basically, this is because these models are linear in the state variable (after a suitable transformation of variables).

OLNEs are time consistent: along the equilibrium path, no player has at any stage any incentive to deviate from its original plan.[3] However, if perhaps by error someone has deviated from the equilibrium path, then the state of the system is revealed to be different from what was predicted, and it will be no longer optimal for a player to continue with its originally planned time path of actions. Thus, an OLNE is not robust to perturbations. In contrast, an MPNE, which consists of Markov-perfect strategies that are best replies to each other, is robust to deviations.[4] A Markov-perfect strategy is a rule that conditions action at any date on the observed state of the system at that date, such that the objective function of the player, starting from any (date, state) pair, is maximized, given the Markov-perfect strategies of the other players.[5] This equilibrium concept corresponds to the idea of subgame perfection.[6]

An OLNE is founded on the assumption that each players has the ability to make a credible precommitment. (This will become clearer in the examples that follow.) The emerging consensus is that in analyzing dynamic games, one should try where possible to find an equilibrium in Markov-perfect strategies.

Reinganum and Stokey (1985) offer an intuitive, nontechnical, explanation of the distinction between open-loop strategies and Markov-perfect strategies. They use the expressions "path strategies" and "decision rule strategies".[7] If agents use path strategies, at the initial date each player must make a binding commitment about the actions it will take at all future dates. In contrast, when a player uses a decision rule, the action at any future date t will depend on the observed value of the state variable at t, denoted by $S(t)$. Decision rules that specify action at time t as a function of the observed pair $(t, S(t))$ only are called Markovian decision rules.[8] As Reinganum and Stokey (1985) point out, a Nash equilibrium in decision rules is not necessarily Markov-perfect. To be Markov-perfect, a Nash equilibrium in decision rules must satisfy the additional property that the continuation of the given decision rules constitutes a Nash equilibrium

---

[3]For a discussion of this point, see Dockner *et al.* (2000).
[4]For a brief discussion of Markov-perfect strategies, see Maskin and Tirole (1988a, p. 553).
[5]For a precise definition, see Dockner *et al.*, 2000.
[6]Of course, in continuous time, the concept of a subgame is problematic.
[7]See also Kydland (1975).
[8]Trigger strategies are examples of decision rules that are not Markovian. See Dockner *et al.* (2000) for a discussion of trigger strategies.

when viewed from *any* future (date, state) pair. Dockner *et al.* (2000, Ex. 4.2) give an example of a Nash equilibrium in decision rules that fails to be Markov-perfect.

Economic models that make use of the equilibrium in feedback strategies are increasingly popular. Beginning with Clemhout *et al.* (1973), Simaan and Cruz (1975), and Levhari and Mirman (1980), more and more dynamic game models focus exclusively on this equilibrium concept.

The economic profession's increasing preference of MPNE over OLNE does not mean that the latter is a useless concept. It is often quite useful to characterize an OLNE and compare it with an MPNE. As Fudenberg and Levine (1988) and Fudenberg and Tirole (2000) point out, an OLNE provides a useful benchmark for understanding the added strategic effect of Markov-perfect strategies.

1.2.1. *A simple transboundary pollution game in discrete time*

To illustrate the difference between OLNE and MPNE, let us consider a simple example: a two-period model of transboundary pollution.

There are two players (say two countries), called SMALL and CAP. Player SMALL chooses its level of CO_2 emissions for periods 1 and 2, denoted by the lowercase symbols x and y. Player CAP's levels of emissions for period 1 and 2 are denoted by the capital letters X and Y. Let S_t be the global stock of CO_2 concentration in the atmosphere at the beginning of period t. The initial stock (at the beginning of period 1) is denoted by S_1. Assume that the stock at the beginning of period 2 is determined by

$$S_2 = S_1 + x + X$$

and that the stock at the end of period 2 (which is the beginning of period 3) is

$$S_3 = S_2 + y + Y.$$

The variable S_t is called the state variable of the system. The control variables in this model are the emission levels.

The period 2 payoff to SMALL is the benefit from consuming the output, $ay - (1/2)y^2$, minus the cost of period 2 environmental damages, $(1/2)S_2^2$:

$$u_2 = ay - \frac{1}{2}y^2 - \frac{1}{2}S_2^2,$$

where a is a positive constant. Similarly, CAP's payoff in period 2 is

$$U_2 = AY - \frac{1}{2}Y^2 - \frac{1}{2}S_2^2,$$

where A is a positive constant.

We assume SMALL's payoff in period 1 is

$$u_1 = ax - \frac{1}{2}x^2 - \frac{1}{2}S_1^2$$

and that of CAP is

$$U_1 = AX - \frac{1}{2}X^2 - \frac{1}{2}S_1^2.$$

The objective of SMALL is to maximize its overall payoff w, defined as the sum of its period 1 and period 2 payoffs, minus a term that reflects its guilt of passing on to the next generation the stock of pollution S_3. We denote this term by $g_3(S_3)$ and assume it is increasing in S_3. For simplicity, let $g_3(S_3)$ take the simple form $(1/2)S_3^2$. Thus, $w \equiv u_1 + u_2 - g_3 = u_1 + u_2 - (1/2)S_3^2$. Similarly, CAP's guilt function is $G_3(S_3) = (1/2)S_3^2$ and it wants to maximize $W \equiv U_1 + U_2 - G_3 = U_1 + U_2 - (1/2)S_3^2$. Let us find the OLNE for this game.

Finding the OLNE

An open-loop strategy of a player is a planned time path of actions over the time horizon. Each player assumes that the other player is going to carry out its planned course of actions.

SMALL, taking as given the time path of actions of CAP denoted by $\{X, Y\}$, chooses its time path $\{x, y\}$ to maximize w:

$$w = ax - \frac{1}{2}x^2 - \frac{1}{2}S_1^2 + ay - \frac{1}{2}y^2 - \frac{1}{2}S_2^2 - \frac{1}{2}S_3^2,$$

where

$$S_2 = S_1 + x + X \quad \text{and} \quad S_3 = S_2 + y + Y = S_1 + x + X + y + Y$$

The first-order conditions are

$$\frac{\partial w}{\partial x} = a - x - (S_1 + x + X) - (S_1 + x + y + X + Y) = 0$$

$$\frac{\partial w}{\partial y} = a - y - (S_1 + x + y + X + Y) = 0$$

CAP, taking $\{x, y\}$ as given, chooses $\{X, Y\}$ to maximize W. The first-order conditions are

$$\frac{\partial W}{\partial X} = A - X - (S_1 + x + X) - (S_1 + x + y + X + Y) = 0$$

$$\frac{\partial W}{\partial Y} = A - Y - (S_1 + x + y + X + Y) = 0$$

Solving these four first-order conditions simultaneously, we obtain the OLNE. Let the superscript OL denotes open-loop strategies. The equilibrium open-loop strategy of SMALL is $\{x^{OL}, y^{OL}\}$, where

$$x^{OL} = \frac{6}{11}a - \frac{4}{11}S_1 - \frac{5}{11}A$$

$$y^{OL} = \frac{7}{11}a - \frac{1}{11}S_1 - \frac{4}{11}A$$

Similarly, that of CAP is $\{X^{OL}, Y^{OL}\}$, where

$$X^{OL} = \frac{6}{11}A - \frac{4}{11}S_1 - \frac{5}{11}a$$

$$Y^{OL} = \frac{7}{11}A - \frac{1}{11}S_1 - \frac{4}{11}a$$

It follows that the time path of the state variable under the OLNE is $\{S_1, S_2^{OL}, S_3^{OL}\}$ where

$$S_2^{OL} = S_1 + x^{OL} + X^{OL} = \frac{1}{11}(3S_1 + a + A)$$

$$S_3^{OL} = S_2 + y^{OL} + Y^{OL} = \frac{1}{11}(S_1 + 4A + 4a)$$

The two-period welfare of SMALL in the OLNE is

$$w = ax^{OL} - \frac{1}{2}(x^{OL})^2 - \frac{1}{2}S_1^2 + ay^{OL} - \frac{1}{2}(y^{OL})^2$$

$$- \frac{1}{2}\left[\frac{1}{11}(3S_1 + a + A)\right]^2 - \frac{1}{2}\left[\frac{1}{11}(S_1 + 4A + 4a)\right]^2.$$

Similarly, the two-period welfare of CAP in the OLNE can be computed.

The OLNE is *time-consistent* in the following sense. Given that in period 1 both players have carried out their respective actions x^{OL} and X^{OL}, at the beginning of period 2, if both players are given an opportunity to revise their plan to maximize the remaining part of their payoffs,

$u_2 - g_3 = (ay - \frac{1}{2}y^2 - \frac{1}{2}S_2^2) - \frac{1}{2}S_3^2$ and $U_2 - G_3 = (AY - \frac{1}{2}Y^2 - \frac{1}{2}S_2^2) - \frac{1}{2}S_3^2$, they would choose respectively the same y^{OL} and Y^{OL}.

However, suppose for some reason (perhaps by error), CAP did not emit the amount X^{OL} in period 1. Say, for example, CAP's emission was $X^{OL} + \varepsilon$. Then, at the beginning of period 2, both players observe that the stock is

$$S_2 = S_2^{OL} + \varepsilon = \frac{1}{11}(3S_1 + a + A) + \varepsilon$$

Will it remain optimal for SMALL to choose $y = 7/11a - 1/11S_1 - 4/11A$? Recall that

$$u_2 - g_3 = \left(ay - \frac{1}{2}y^2 - \frac{1}{2}S_2^2\right) - \frac{1}{2}S_3^2$$

$$= ay - \frac{1}{2}y^2 - \frac{1}{2}\left[\frac{1}{11}(3S_1 + a + A) + \varepsilon\right]^2$$

$$- \frac{(\frac{1}{11}(3S_1 + a + A) + \varepsilon + y + Y)^2}{2}$$

The first-order condition for SMALL's optimal choice y is

$$\frac{\partial(u_2 - g_3)}{\partial y} = a - y - \left(\frac{1}{11}(3S_1 + a + A) + \varepsilon + y + Y\right) = 0$$

Similarly for CAP

$$\frac{\partial(U_2 - G_3)}{\partial Y} = A - Y - \left(\frac{1}{11}(3S_1 + a + A) + \varepsilon + y + Y\right) = 0$$

Because of the presence of the perturbation ε, these two first-order conditions yield emissions (y, Y) that are not the same as the originally planned quantities (y^{OL}, Y^{OL}). This shows that the OLNE is not robust to deviation, that is, it is not subgame perfect.

When the second period comes, a player would be foolish to stick to the previously planned emission Y^{OL} if the observed pollution stock S_2 turned out to be different from S_2^{OL}. It would seem, then, that in an environment where perturbations are possible, each player would be wise to think that the other player would act in each period according to the observed level of stock. This gives rise to the idea of a feedback (or Markov-perfect) strategy: optimal current action should depend on currently observed state.

Finding the MPNE

Markov-perfect strategies are found by solving the game backward. This method gives us "feedback decision rules" (e.g., emission in a given period depends on the observed stock at the beginning of that period).

At the beginning of period 2, given the observed stock level S_2, SMALL chooses y to maximize

$$u_2 - g_3 = ay - \frac{1}{2}y^2 - \frac{1}{2}S_2^2 - \frac{1}{2}(S_2 + y + Y)^2$$

The first-order condition is

$$a - y - (S_2 + y + Y) = 0$$

Thus, we obtain SMALL's reaction function for period 2:

$$y = \frac{a - Y - S_2}{2}$$

Similarly, CAP's reaction function in period 2 is

$$Y = \frac{A - y - S_2}{2}$$

The intersection of the two reaction curves determines the period-two Nash equilibrium emissions, expressed as functions of the observed stock S_2. Thus, we get the feedback (FB) decision rules for SMALL and CAP

$$y^{\text{FB}}(S_2) = \frac{2}{3}a - \frac{1}{3}S_2 - \frac{1}{3}A$$

$$Y^{\text{FB}}(S_2) = \frac{2}{3}A - \frac{1}{3}S_2 - \frac{1}{3}a,$$

where the superscript FB indicates that this is a feedback equilibrium. Note that these decision rules indicate that a player will pollute less in period 2 if it observes a greater level of pollution at the beginning of that period.[9]

The resulting feedback equilibrium stock S_3 is then

$$S_3 = S_2 + y^{\text{FB}}(S_2) + Y^{\text{FB}}(S_2) = \frac{1}{3}(S_2 + A + a).$$

Let us work backward to find the equilibrium decision rules for period 1. The feedback-equilibrium payoff to SMALL for period 2 (including the terminal

---

[9]As we show below, knowing this, each player *has an incentive to pollute a bit more in period 1* (to increase its first period payoff), as it believes the other player will reduce its own emissions in period 2.

term $g_3(S_3)$) is

$$u_2 - g_3 = ay^{\text{FB}}(S_2) - \frac{1}{2}\left(y^{\text{FB}}(S_2)\right)^2 - \frac{1}{2}S_2^2$$
$$- \frac{1}{2}(S_2 + y^{\text{FB}}(S_2) + Y^{\text{FB}}(S_2))^2 \equiv v_2(S_2).$$

Similarly

$$U_2 = AY^{\text{FB}}(S_2) - \frac{1}{2}\left(Y^{\text{FB}}(S_2)\right)^2 - \frac{1}{2}S_2^2 - \frac{1}{2}\left(S_2 + y^{\text{FB}}(S_2) + Y^{\text{FB}}(S_2)\right)^2$$
$$\equiv V_2(S_2).$$

Working backward, at the beginning of period 1, SMALL chooses x to maximize

$$u_1 + v_2(S_2) \equiv ax - \frac{1}{2}x^2 - \frac{1}{2}S_1^2 + v_2(S_2),$$

where $S_2 = S_1 + x + X$.

SMALL's first-order condition is

$$\frac{\partial u_1}{\partial x} + \frac{dv_2}{dS_2}\frac{\partial S_2}{\partial x} = 0,$$

where

$$\frac{dv_2}{dS_2} = \frac{\partial(u_2 - g_3)}{\partial y^{\text{FB}}}\frac{dy^{\text{FB}}}{dS_2} + \frac{\partial(u_2 - g_3)}{\partial Y^{\text{FB}}}\frac{dY^{\text{FB}}}{dS_2} + \frac{\partial(u_2 - g_3)}{\partial S_2}$$

$$= 0 - S_3\frac{dY^{\text{FB}}}{dS_2} - S_2 - S_3$$

$$= -S_2 - \frac{1}{3}(S_2 + A + a)\left[1 + \frac{dY^{\text{FB}}}{dS_2}\right] = -S_2 - \frac{2}{9}(S_2 + A + a)$$

$$= -\frac{(11S_2 + 2A + 2a)}{9}.$$

Therefore, SMALL's first-order condition is simply

$$a - x - \frac{(11S_2 + 2A + 2a)}{9} = 0.$$

That is,

$$a - x - \frac{(11S_1 + 11x + 11X + 2A + 2a)}{9} = 0.$$

This equation yields SMALL's period 1 reaction function:

$$x = \frac{-11S_1 - 11X - 2A + 7a}{20}$$

Similarly, CAP's period 1 reaction function is

$$X = \frac{-11S_1 - 11x - 2a + 7A}{20}$$

The intersection of these two reaction curves gives the equilibrium feedback decision rules in period 1:

$$x^{FB}(S_1) = \frac{18}{31}a - \frac{11}{31}S_1 - \frac{13}{31}A$$

$$X^{FB}(S_1) = \frac{18}{31}A - \frac{11}{31}S_1 - \frac{13}{31}a.$$

The two-period welfare of each player can then be computed, and expressed as functions of the parameters of the model (in our example, the parameters are S_1, a, and A).

Comparing OLNE and MPNE of the transboundary pollution game

Let us compare the first-period emissions under OLNE with those under MPNE. For SMALL,

$$x^{OL} = \frac{6}{11}a - \frac{4}{11}S_1 - \frac{5}{11}A$$

$$x^{FB}(S_1) = \frac{18}{31}a - \frac{11}{31}S_1 - \frac{13}{31}A$$

Both equations indicate that a higher initial stock level S_1 will cause players to opt for lower emissions in period 1, but since $4/11 > 11/31$, we can see that players are more willing to cut back their emissions under OLNE than under MPNE. On the other hand, an exogenous increase in the preference parameter a will lead to more emissions, but this effect is much more pronounced under the MPNE.

An interesting question is: does the OLNE give rise to a higher overall payoff for each player than the MPNE? If it does, one could argue that in the context of this model, *institutions that facilitate precommitment would be welfare enhancing.*

Let us consider a numerical example. Let $A = a = 20$, and $S_1 = 2$. Then, $x^{OL} = X^{OL} = 1.09$, $y^{OL} = Y^{OL} = 5.27$, $S_2^{OL} = 4.18$ and $S_3^{OL} = 14.72$. Thus, the overall payoff of each player in the OLNE is equal to -6.41.

Compare with the MPNE: $x^{\mathrm{FB}} = X^{\mathrm{FB}} = 2.51 > 1.09$. This results in $S_2 = 7.03$. The second-period emissions are $y^{\mathrm{FB}} = Y^{\mathrm{FB}} = 4.32 < 5.27$. The resulting S_3 is 15.67. The overall payoff of each under the MPNE is -25.34.

The above numerical example indicates that both players are better off if they both use open-loop strategies, under which, given S_1, each player would commit to a time path of actions, regardless of what S_2 turns out to be. However, in the absence of a mechanism to ensure that they honor their commitments, each player would believe that the other player will deviate from the committed action for period 2 if S_2 turns out to be different from the level implied by their committed period 1 emissions. Suppose CAP believes this. Then, it will deviate from the OLNE by increasing its period 1 emissions beyond its OLNE level, so as to increase S_2, knowing that SMALL will then reduce its own period 2 emissions, as dictated by the rule $y^{\mathrm{FB}}(S_2)$ that we discovered above. Thus, CAP's first-period deviation from OLNE will increase its overall payoff (because it manages to pollute more while getting SMALL to pollute less in period 2). If SMALL anticipates this, it will also deviate from first period OLNE emissions.

The above discussion indicates that while an OLNE might achieve higher overall payoff for both players, their mutual suspicion and opportunistic behavior will prevent an OLNE from being realized.

1.2.2. *Choice among equilibrium concepts*

When one formulates and analyzes a dynamic game, should one focus only on MPNE? Is the MPNE likely to be a better prediction of the outcome of the game?

Recall that in an MPNE, each player assumes that the other player will act in each period according to the observed level of the state variable in that period. Each player therefore has an incentive to influence the state variable in period t with the objective of manipulating the other player's action in period $t + 1$. Thus, MPNE is perhaps a better concept if one believes that players are sophisticated and manipulative.

If players realize that, in a given game, the OLNE gives higher welfare to each than the MPNE does, would there be incentives for them to agree on playing open-loop strategies? Such an agreement would require an ability to commit to an initially announced time path of actions.

One may argue that in situations where agents are able to commit, an OLNE may well be a good prediction of the outcome of the game. Another advantage of OLNE is that it is relatively easier to compute than MPNE.

In situations where calculations are extremely costly, there is a plausible presumption that economic agents may opt for the easy-to-compute OLNE. In addition, there are situations where an OLNE is so plausible and intuitive that it is not worthwhile to try to compute MPNE for a small gain in sophistication. We later provide an example of this type.

It is worth noting that OLNE and MPNE can be thought of as based on two alternative, both extreme, assumptions about ability to precommit. In the OLNE, players commit to a whole time path of play. In the MPNE, players cannot precommit at all. Reinganum and Stokey (1985) argue that in some cases, players may be able to commit to actions in the near future (e.g., by forward contracts) but not to actions in the distant future. They develop a simple model where the game begins at time 0 and ends at a fixed time T, and there are k periods of equal length δ, where $k\delta = T$. At the beginning of each period, agents can commit to a path of action during that period. The special case where $k = 1$ corresponds to the open-loop formulation, and the OLNE is then the appropriate equilibrium concept. At the other extreme, where $\delta \to 0$, the appropriate equilibrium concept is MPNE. This issue is discussed further in the chapter on natural resources (Chap. 3).

We now turn to a simple example where there is a very plausible OLNE.

1.2.3. *Another simple dynamic game: non-cooperative cake eating*

Two players must share a cake that has been cut into six identical pieces. They have three days, $t = 1, 2, 3$, to play this game. Let s_t denote the number of pieces of cake that remain at the beginning of day t. Then, $s_1 = 6$. Each day t, after observing the number of pieces of cake that remain, s_t, each player must touch one of the three buttons marked $0, 1$, and 2 on its computer screen. To put it more formally, the control variable for player i, denoted by b_{it}, can take one of the following three values $0, 1$, and 2. According to the rules of the game, if $b_{1t} + b_{2t} \leq s_t$ then player i ($i = 1, 2$) is given b_{it} pieces and consumes them. Its utility for that day is the square root of the number of pieces it consumes:

$$u_{it} = \sqrt{b_{it}}$$

If $b_{1t} + b_{2t} > s_t$, then both players get nothing, and their utility is 0.

The three-period welfare of player i $(i = 1, 2)$ is assumed to be the non-discounted sum of utilities:

$$w_i = u_{i1} + u_{i2} + u_{i3}$$

It is easy to see that the following pair of open-loop strategies constitutes an OLNE: each player chooses 1 each day. Along the equilibrium play, each will consume one piece each day. Its equilibrium three-period welfare is then

$$\sqrt{1} + \sqrt{1} + \sqrt{1} = 3$$

It can be verified that it does not pay any player to deviate from this equilibrium. Suppose player 2 uses the open-loop equilibrium strategy ($b_{2t} = 1$ for $t = 1, 2, 3$) and player 1 deviates, say, by choosing $b_{11} = b_{12} = 2$ and $b_{13} = 2$ (or 1) then player 1's payoff will be

$$\sqrt{2} + \sqrt{2} + \sqrt{0} = 2.8284 < 3$$

Therefore, this deviation is not profitable. It is easy to check that in fact there is no profitable deviation.

It is interesting to observe that the OLNE of this game is Pareto optimal. If the two players were allowed to collude in their choice of strategies, they would not achieve better outcome.

Let us show that the OLNE is not subgame perfect. Suppose both players have decided to play the OLNE strategies ($b_{it} = 1$ for $t = 1, 2, 3$ for $i = 1, 2$) but, on the second day, player 1 observes that there are only three pieces of cake left (i.e., at least one player has made a mistake by touching the wrong button). Would player 1 still want to continue with its original open-loop strategy? If it does (and assuming player 2 will play $b_{2t} = 1$ for $t = 2, 3$), it will get

$$\sqrt{1} + \sqrt{1} + \sqrt{0} = 2$$

If it chooses instead $b_{12} = 2$, it will get a better payoff

$$\sqrt{1} + \sqrt{2} + \sqrt{0} = 2.4142$$

assuming player 2 will play $b_{2t} = 1$ for $t = 2, 3$. This shows that the OLNE is, in general, not subgame perfect: if the time path of the state variable is observed to have deviated from the equilibrium path, at least one player will want to deviate from the original plan.

The reader is invited to find all the MPNEs for this game (recall that an MPNE is a pair of feedback strategies that are best replies to each other) and to show that the outcome of any MPNE of this game is that each player consumes one piece of cake each day, that is, the same outcome as that obtained under OLNE.

1.2.4. *An infinite horizon game of transboundary pollution in continuous time*

Let us consider the following game proposed by Long (1992). There are two countries. Country i's output at date t is denoted by $y_i(t)$. Assume all output is consumed. Emissions are proportional to output, $E_i(t) = y_i(t)$ where the factor of proportionality is normalized at unity. (In what follows, we use y_i and E_i interchangeably.) The stock of pollution, common to both countries, is $S(t)$. The rate of accumulation of the stock is equal to the sum of emissions minus the natural decay:

$$\dot{S}(t) = E_1(t) + E_2(t) - \delta S(t), \tag{1.1}$$

where $\delta > 0$ is the decay rate. The pollution damage suffered by country i at time t is

$$D_i(S(t)) = \frac{c_i}{2}(S(t))^2,$$

where $c_i > 0$ is the damage parameter. The utility of consumption is $U_i(y_i(t)) = A_i y_i(t) - 1/2(y_i(t))^2$, where A_i is a positive constant. The net utility, denoted by $B_i(t)$, is defined as the utility of consumption minus the damage cost:

$$B_i(t) = A_i y_i(t) - \frac{1}{2}(y_i(t))^2 - \frac{c_i}{2}(S(t))^2$$

The government of country i perceives that the country's social welfare is

$$W_i = \int_0^\infty e^{-\rho t} B_i(t) dt,$$

where $\rho > 0$ is the rate of discount. Its objective is to maximize the country's social welfare subject to the transition Eq. (1.1). In doing so, it must know if the government of the other country uses an open-loop or a feedback-emission strategy. These two cases are examined separately below.

OLNE in the infinite horizon model

If country i believes that country j uses an open-loop emission strategy, $E_j(t) = \phi_j^{OL}(t)$, its optimization problem becomes

$$\max_{E_i(.)} \int_0^\infty e^{-\rho t} \left[A_i E_i(t) - \frac{1}{2}(E_i(t))^2 - \frac{c_i}{2}(S(t))^2 \right] dt \qquad (1.2)$$

subject to

$$\dot{S}(t) = E_i(t) + \phi_j^{OL}(t) - \delta S(t), \qquad S(0) = S_0. \qquad (1.3)$$

This is a simple optimal control problem, where $\phi_j^{OL}(t)$ is taken as an exogenously given time path. Let us solve this problem using the Maximum Principle.[10] Let H_i denote the Hamiltonian function for the optimal control problem (1.2), and ψ_i be the co-state variable. Then

$$H_i = A_i E_i - \frac{1}{2}(E_i)^2 - \frac{c_i}{2}(S)^2 + \psi_i(E_i + \phi_j^{OL} - \delta S)$$

The necessary conditions are

$$\frac{\partial H_i}{\partial E_i} = A_i - E_i + \psi_i = 0 \qquad (1.4)$$

$$-(\dot{\psi}_i - \rho \psi_i) = \frac{\partial H_i}{\partial S} = -c_i S - \psi_i \delta \qquad (1.5)$$

$$\dot{S} = \frac{\partial H_i}{\partial \psi_i} = E_i + \phi_j^{OL} - \delta S \qquad (1.6)$$

and the transversality condition is

$$\lim_{t \to \infty} e^{-\rho t} \psi_i(t) S(t) = 0 \qquad (1.7)$$

The variable ψ_i can be eliminated by using the necessary condition (1.4), and thus we get the following conditions

$$-(\dot{E}_i - \rho(E_i - A_i)) = -c_i S - \delta(E_i - A_i)$$

$$\dot{S} = E_i + \phi_j^{OL} - \delta S, \qquad S(0) = S_0$$

$$\lim_{t \to \infty} e^{-\rho t}(E_i(t) - A_i) S(t) = 0.$$

---

[10]See, for example, Kamien and Schwartz (1991) or Léonard and Long (1992).

A similar set of equations applies to country j's optimization problem, if j believes that i uses an open-loop emissions strategy $E_i(t) = \phi_i^{\mathrm{OL}}(t)$:

$$-(\dot{E}_j - \rho(E_j - A_j)) = -c_j S - \delta(E_j - A_j)$$

$$\dot{S} = E_j + \phi_i^{\mathrm{OL}} - \delta S$$

$$\lim_{t \to \infty} e^{-\rho t}(E_j(t) - A_j)S(t) = 0$$

To find an OLNE, we must find a pair of functions $(\phi_1^{\mathrm{OL}}, \phi_2^{\mathrm{OL}})$ such that $\phi_1^{\mathrm{OL}}(t) = E_1^*(t)$ and $\phi_2^{\mathrm{OL}}(t) = E_2^*(t)$ where $(E_1^*(t), E_2^*(t), S^*(t))$ satisfy the three differential equations

$$\dot{E}_1(t) = c_1 S(t) + (\rho + \delta)(E_1(t) - A_1) \tag{1.8}$$

$$\dot{E}_2(t) = c_2 S(t) + (\rho + \delta)(E_2(t) - A_2) \tag{1.9}$$

$$\dot{S}(t) = E_1(t) + E_2(t) - \delta S(t), \qquad S(0) = S_0 \tag{1.10}$$

and, in addition, the following transversality conditions

$$\lim_{t \to \infty} e^{-\rho t}(E_1(t) - A_1)S(t) = 0 \tag{1.11}$$

$$\lim_{t \to \infty} e^{-\rho t}(E_2(t) - A_2)S(t) = 0 \tag{1.12}$$

This is a system of three differential equations with three boundary conditions. Before solving for an OLNE for this system let us consider the special case where the cost and preference parameters of the two countries are identical, that is, $A_1 = A_2 = A$ and $c_1 = c_2 = c$. In this case, it is reasonable to assume that the two countries will adopt identical open-loop strategies, that is, $E_1^*(t) = E_2^*(t) = E^*(t)$. Then, the system reduces to two differential equations

$$\dot{E}(t) = cS(t) + (\rho + \delta)(E(t) - A) \tag{1.13}$$

$$\dot{S}^*(t) = 2E(t) - \delta S(t), \qquad S(0) = S_0 \tag{1.14}$$

with the transversality condition

$$\lim_{t \to \infty} e^{-\rho t}(E(t) - A)S(t) = 0. \tag{1.15}$$

The pair of differential Eqs. (1.13) and (1.14) admits a unique steady-state pair $(\widehat{S}, \widehat{E})$, where

$$\widehat{S} = \frac{2A(\delta + \rho)}{2c + \delta(\delta + \rho)} \tag{1.16}$$

$$\widehat{E} = \frac{A\delta(\delta + \rho)}{2c + \delta(\delta + \rho)} = \frac{\delta \widehat{S}}{2}, \tag{1.17}$$

Remark 1 If the two countries co-operate and maximize the sum of their welfares, the steady-state stock of pollution will be lower than $\widehat{S}$.

We can show that steady-state pair $(\widehat{S}, \widehat{E})$ has the saddle-point property, in the sense that for any given S_0, there exists a *unique* corresponding $E^*(0)$ such that if both countries choose $E(0)$ as their initial emission rate, the pair of time paths $(S^*(t), E^*(t))$ starting with $(S_0, E^*(0))$ at time zero will converge to the steady-state pair $(\widehat{S}, \widehat{E})$. Of course, this pair of time paths satisfies the transversality condition (1.15). We show below how to determine $E^*(0)$, given S_0.

To prove the saddle-point property, we must show that the Jacobian matrix J of the system (1.13)–(1.14) has exactly one negative real root, that is, it satisfies the condition $\det J < 0$ (recall that the product of the roots equals the determinant of J). Now

$$J = \begin{bmatrix} \rho + \delta & c \\ 2 & -\delta \end{bmatrix} \equiv \begin{bmatrix} a_{11} & a_{12} \\ a_{21} & a_{22} \end{bmatrix}$$

$$\det J = a_{11}a_{22} - a_{12}a_{21} = -\delta(\rho + \delta) - 2c < 0.$$

Recall that the trace of matrix J is $\text{trace}(J) = a_{11} + a_{12}$. The characteristic equation is

$$\lambda^2 - \text{trace}(J)\lambda + \det J = 0$$

or

$$\lambda^2 - \rho\lambda - \delta(\rho + \delta) - 2c = 0. \tag{1.18}$$

Equation (1.18) has one negative real root, denoted by λ_1, and one positive real root, denoted by λ_2:

$$\lambda_1 = \frac{\text{trace}(J) - \sqrt{(\text{trace}(J))^2 - 4\det J}}{2}$$

$$= \frac{\rho - \sqrt{(\rho + 2\delta)^2 + 8c}}{2} < 0 \tag{1.19}$$

$$\lambda_2 = \frac{\rho + \sqrt{(\rho + 2\delta)^2 + 8c}}{2} > 0 \tag{1.20}$$

We take the negative root for convergence. The convergent paths of E and S (along the stable branch of the saddle point) are given by

$$\begin{bmatrix} E^*(t) - \widehat{E} \\ S^*(t) - \widehat{S} \end{bmatrix} = \beta \begin{bmatrix} \mathbf{k}_1^{(1)} \\ \mathbf{k}_2^{(1)} \end{bmatrix} e^{\lambda_1 t} \equiv \beta \mathbf{k} e^{\lambda_1 t}, \tag{1.21}$$

where $\mathbf{k}$ is a characteristic vector corresponding to the negative root λ_1, and β is a constant. Take

$$\mathbf{k} = \begin{bmatrix} \mathbf{k}_1^{(1)} \\ \mathbf{k}_2^{(1)} \end{bmatrix} = \begin{bmatrix} 1 \\ (\lambda_1 - a_{11})a_{12}^{-1} \end{bmatrix}.$$

Then, Eq. (1.21) gives

$$\frac{E^*(t) - \widehat{E}}{S^*(t) - \widehat{S}} = \frac{1}{(\lambda_1 - a_{11})a_{12}^{-1}} \tag{1.22}$$

Therefore, the right-hand side (RHS) of Eq. (1.22) is the slope of the stable branch of the saddle point in the space (S, E). At $t = 0$, $S^*(0) = S_0$ (given). It follows that given S_0, the initial emission rate dictated by the OLNE is

$$E^*(0) = \widehat{E} + \frac{S_0 - \widehat{S}}{(\lambda_1 - a_{11})a_{12}^{-1}} = \widehat{E} + \frac{(S_0 - \widehat{S})c}{\lambda_1 - \rho - \delta} \tag{1.23}$$

Remark 2   The matrix Eq. (1.21) gives

$$E^*(t) - \widehat{E} = \beta \mathbf{k}_1^{(1)} e^{\lambda_1 t} \tag{1.24}$$

$$S^*(t) - \widehat{S} = \beta \mathbf{k}_1^{(2)} e^{\lambda_1 t}. \tag{1.25}$$

Differentiating these equations with respect to t, we get, after substitution,

$$\dot{E}^*(t) = \lambda_1 (E^*(t) - \widehat{E}) \tag{1.26}$$

$$\dot{S}^*(t) = \lambda_1 (S^*(t) - \widehat{S}). \tag{1.27}$$

The entire time path of E^* can be obtained using the first-order differential Eq. (1.26) and the initial value $E^*(0)$ as given by Eq. (1.23). Thus

$$E^*(t) = \widehat{E} + \left(E^*(0) - \widehat{E} \right) e^{\lambda_1 t} \tag{1.28}$$

Similarly

$$S^*(t) = \widehat{S} + (S_0 - \widehat{S}) e^{\lambda_1 t} \tag{1.29}$$

Remark 3 If initially the environment is perfectly clean, that is, if $S_0 = 0$, then the (OLNE) initial emission rate is

$$E_0^* = \widehat{E} - \frac{\widehat{S}}{(\lambda_1 - a_{11})a_{12}^{-1}} = \widehat{S}\left[\frac{\delta}{2} - \frac{1}{(\lambda_1 - a_{11})a_{12}^{-1}}\right] \qquad (1.30)$$

Remark 4 An alternative representation of the open-loop equilibrium path is a diagram in the $(S, \dot{S})$ space. Equation (1.27) describes the movement of $S^*(t)$ that results from the OLNE. It can be depicted in a simple diagram in the $(S, \dot{S})$ space.

Remark 5 An alternative method of finding $E^*(0)$ is as follows. At time $t = 0$, Eq. (1.27) gives

$$2E^*(0) - \delta S(0) = \lambda_1(S(0) - \widehat{S}), \qquad (1.31)$$

which determines $E^*(0)$ uniquely. Note that in the special case where initially the environment is perfectly clean, that is, if $S_0 = 0$, Eq. (1.31) gives

$$2E_0^* = \lambda_1 \widehat{S} \qquad (1.32)$$

Equation (1.32) is consistent with Eq. (1.30) because $2a_{12} = (\lambda_1 + \delta)(\lambda_1 - a_{11})$.

Remark 6 From Eq. (1.27), we get

$$2E^*(t) - \delta S^*(t) = \lambda_1 \left[S^*(t) - \widehat{S}\right]$$

Hence, along the open-loop equilibrium play, there is a linear relationship between E^* and S^*:

$$E^* = \frac{1}{2}\left[(\lambda_1 + \delta)S^* - \lambda \widehat{S}\right] \qquad (1.33)$$

This equation is sometimes called the "feedback representation of the OLNE". It should not be interpreted as a feedback strategy.

Let us return to the general case where the two countries differ, that is, $A_1 \neq A_2$ and $c_1 \neq c_2$. We must solve for the OLNE using the system of three differential Eq. (1.8)–(1.10). We show that there exists a unique solution $(E_1^*(t), E_2^*(t), S^*(t))$ that converges to a unique steady state $(\widehat{E}_1, \widehat{E}_2, \widehat{S}^a)$. Here, the superscript a in $\widehat{S}^a$ indicates that we are dealing with the case of an asymmetric world. The vector $(\widehat{E}_1, \widehat{E}_2, \widehat{S}^a)$ is the solution of the matrix equation

$$\begin{bmatrix} \rho + \delta & 0 & c_1 \\ 0 & \rho + \delta & c_2 \\ 1 & 1 & -\delta \end{bmatrix} \begin{bmatrix} \widehat{E}_1 \\ \widehat{E}_2 \\ \widehat{S}^a \end{bmatrix} = \begin{bmatrix} -(\rho + \delta)A_1 \\ -(\rho + \delta)A_2 \\ 0 \end{bmatrix}$$

The steady state $(\widehat{E}_i, \widehat{E}_j, \widehat{S}^a)$ has the saddle-point property. To verify this, consider the Jacobian matrix

$$J_3 \equiv \begin{bmatrix} \rho + \delta & 0 & c_1 \\ 0 & \rho + \delta & c_2 \\ 1 & 1 & -\delta \end{bmatrix}$$

Then

$$\det J_3 = -\delta(\rho + \delta)^2 - (c_1 + c_2)(\rho + \delta) < 0$$

The steady-state values are

$$\begin{bmatrix} \widehat{E}_1 \\ \widehat{E}_2 \\ \widehat{S} \end{bmatrix} = \frac{1}{\det J_3} \begin{bmatrix} -A_1\delta(\rho + \delta)^2 - A_1 c_2(\rho + \delta) + A_2 c_1(\rho + \delta) \\ -A_2\delta(\rho + \delta)^2 - A_2 c_1(\rho + \delta) + A_1 c_2(\rho + \delta) \\ -(A_1 + A_2)(\rho + \delta)^2 \end{bmatrix}$$

Thus

$$\widehat{S}^a = \frac{(A_1 + A_2)(\rho + \delta)}{(c_1 + c_2) + \delta(\rho + \delta)} \tag{1.34}$$

The determinant of J_3 is negative and the trace of J_3 is positive. This implies there is a negative root, and two roots with positive real parts. This shows that the system exhibits saddle-point stability: given S_0, one can determine a unique pair $(E_i^*(0), E_j^*(0))$ such that starting from $(E_i^*(0), E_j^*(0), S_0)$, the path $(E_i^*(t), E_j^*(t), S(t))$ converges to the steady state $(\widehat{E}_i, \widehat{E}_j, \widehat{S}^a)$. The stable branch of the saddle-point is a line in the three-dimensional space (E_i, E_j, S). It is a one-dimensional stable manifold.

Remark 7 We can directly compute the roots of the characteristic equation

$$\det[J_3 - \lambda I] = 0$$

This equation is a cubic equation in λ

$$(\lambda - \delta - \rho)^2(\lambda + \delta) - (\lambda - \delta - \rho)(c_1 + c_2) = 0$$

and the roots are

$$\lambda_1 = \frac{1}{2}\left[\rho - \sqrt{(\rho + 2\delta)^2 + 4(c_1 + c_2)}\right] < 0$$

$$\lambda_2 = \frac{1}{2}\left[\rho + \sqrt{(\rho + 2\delta)^2 + 4(c_1 + c_2)}\right] > 0$$

$$\lambda_3 = \delta + \rho > 0$$

We choose the negative root λ_1 for convergence. Then, we have the solution path for S:

$$S(t) = \widehat{S}^a + (S_0 - \widehat{S}^a)e^{\lambda_1 t}$$

Finding an MPNE in the infinite horizon model

Suppose country i believes that country j uses a feedback emissions strategy, $E_j(t) = \phi_j^{\mathrm{FB}}(S(t))$, that is, the rate of emissions at t is dependent on the currently observed level of the stock, $S(t)$. Then, the optimal control problem for country i is

$$\max_{E_i(\cdot)} \int_0^\infty e^{-\rho t}\left[A_i E_i(t) - \frac{1}{2}(E_i(t))^2 - \frac{c_i}{2}(S(t))^2\right] dt \qquad (1.35)$$

subject to

$$\dot{S}(t) = E_i(t) + \phi_j^{\mathrm{FB}}(S(t)) - \delta S(t), \qquad S(0) = S_0. \qquad (1.36)$$

Notice that $\phi_j^{\mathrm{FB}}(S)$ is a function of S, and not a function of t. Country i therefore knows that if it influences S, it will indirectly influence the emission rate chosen by country j. This adds a strategic consideration, which was not present in the open-loop case. Let us see how this additional strategic consideration affects the necessary conditions for i's optimal control problem. The Hamiltonian for country i is

$$H_i = A_i E_i - \frac{1}{2}(E_i)^2 - \frac{c_i}{2}(S)^2 + \psi_i(E_i + \phi_j^{\mathrm{FB}}(S) - \delta S).$$

The necessary conditions are

$$\frac{\partial H_i}{\partial E_i} = A_i - E_i + \psi_i = 0 \qquad (1.37)$$

$$-(\dot{\psi}_i - \rho\psi_i) = \frac{\partial H_i}{\partial S} = -c_i S + \psi_i \frac{d\phi^{\mathrm{FB}}(S)}{dS} - \psi_i \delta \qquad (1.38)$$

$$\dot{S} = E_i + \phi_j^{\mathrm{FB}}(S) - \delta S, \qquad S(0) = S_0 \qquad (1.39)$$

and the transversality condition is

$$\lim_{t \to \infty} e^{-\rho t} \psi_i(t) S(t) = 0 \tag{1.40}$$

Comparing Eqs. (1.38) and (1.5), we see that in the feedback case, there is an extra term, $\psi_i d\phi^{\mathrm{FB}}(S)/dS$. This term reflects the additional strategic consideration that when i takes an action, it realizes that the action will change the future level of S, which will in turn influence j's emissions.

Let us consider the simplest case, where the two countries have identical preference and cost parameters. We focus on the symmetric equilibrium. Substituting $E - A$ for ψ, we obtain from Eq. (1.38)

$$-(\dot{E} - \rho(E - A)) = -cS + (E - A)\frac{dE(S)}{dS} - (E - A)\delta \tag{1.41}$$

and

$$\dot{S} = 2E(S) - \delta S \tag{1.42}$$

Now, recall that we are looking for a feedback strategy $E = E(S) \equiv \phi^{\mathrm{FB}}(S)$. Thus,

$$\dot{E} = \frac{dE}{dS}\frac{dS}{dt} = (2E(S) - \delta S)\frac{dE}{dS}.$$

Therefore, Eq. (1.41) can be written as

$$-(2E(S) - \delta S)\frac{dE}{dS} + (\rho + \delta)[E(S) - A] = -cS + (E(S) - A)\frac{dE(S)}{dS}. \tag{1.43}$$

Simplify to get

$$\frac{dE}{dS} = \frac{(\rho + \delta)[E(S) - A] + cS}{3E(S) - A - \delta S} \tag{1.44}$$

Equation (1.44) is a first-order differential equation, where S is the independent variable and E is the dependent variable. Does this differential equation, together with the transversality condition (1.40) where $\psi_i(t)$ is replaced by $E(t) - A$, help determine an MPNE? It turns out that a more direct approach, using the Hamilton–Jacobi–Bellman (HJB) equation, would give us a clearer picture.

The HJB equation for country i is

$$\rho V_i(S) = \max_{E_i} \left[AE_i - \frac{1}{2}E_i^2 - \frac{c}{2}S^2 + V_i'(S)(E_i + E_j(S) - \delta S) \right], \quad (1.45)$$

where $E_j(S)$ is country j's feedback strategy, and $V_i(S)$ is country i's value function, to be solved for. We also impose the condition that along the equilibrium path, the value does not grow too fast:[11]

$$\lim_{t \to \infty} e^{-\rho t} V(S(t)) = 0 \quad (1.46)$$

Maximizing the RHS of the HJB equation with respect to E_i gives the FOC

$$A - E_i + V_i'(S) = 0$$

Rearranging terms

$$E_i = A + V_i'(S)$$

This gives $E_i = E_i(S)$, that is, the chosen emission rate at any time t depends only on the observed level of the stock, and is independent of time. Now, since we are assuming that the two countries have identical parameter values, we focus on the symmetric solution, where $V_i'(S) = V_j'(S) = V'(S)$ and $E_i(S) = E_j(S) = E(S)$. Then, substituting E by $A + V'(S)$ into the HJB equation, we obtain

$$\rho V(S) = \frac{1}{2}[A^2 + 4AV' + 3(V')^2] - \delta SV' - \frac{c}{2}S^2 \quad (1.47)$$

This is a first-order differential equation which, together with condition (1.46), helps determine an MPNE.

Remark 8 If we differentiate Eq. (1.47) with respect to S, then re-arrange terms and substitute $V'(S) = E(S) - A$ and $V''(S) = E'(S)$, we obtain an equation identical to Eq. (1.44). This shows that the two approaches are, in fact, equivalent.

---

[11]See Dockner *et al.* (2000, Sec. 3.6) for weaker conditions.

Let us conjecture that the value function is quadratic[12]

$$V(S) = -\frac{\alpha S^2}{2} - \beta S - \mu \qquad (1.48)$$

Then,

$$V'(S) = -\alpha S - \beta \qquad (1.49)$$

and we get the linear feedback strategy

$$E(S) = A - \beta - \alpha S \qquad (1.50)$$

We expect that $\alpha > 0$, that is, a higher stock will lead countries to reduce emissions, and that $\beta > 0$, that is, if $S = 0$, the marginal effect on welfare of an exogenous increase in S is negative.

Substituting Eqs. (1.48) and (1.49) into Eq. (1.47), we obtain a quadratic equation of the form

$$p_0 + p_1 S + p_2 S^2 = 0,$$

where p_0, p_1, and p_2 are expressions involving the parameters δ, ρ, and c and the coefficients α, β, and μ (to be determined). Since this equation must hold for all S, it follows that the following conditions must be satisfied:

$$p_0 = 0$$

$$p_1 = 0$$

$$p_2 = 0,$$

where the last condition implies that

$$p_2 \equiv \frac{3}{2}\alpha^2 + \left(\delta + \frac{\rho}{2}\right)\alpha - \frac{c}{2} = 0 \qquad (1.51)$$

These conditions yield the following values for α, β, and μ:

$$\alpha = \frac{1}{3}\left[-\left(\delta + \frac{\rho}{2}\right) + \sqrt{\left(\delta + \frac{\rho}{2}\right)^2 + 3c} \right] \equiv \alpha_m \qquad (1.52)$$

---

[12]The quadratic value function gives rise to linear strategies. Dockner and Long (1993) show that for this model, there exist non-linear strategies as well, in which case the corresponding value function is not quadratic. This possibility was pointed out by Reynold (1987), Tsutsui and Mino (1990), and further discussed in Clemhout and Wan (1994). The non-uniqueness is due to the lack of a natural boundary condition, that is, the assumption of feedback strategies is not restrictive enough to yield a unique steady-state and unique feedback strategies.

(We show below that for convergence to a steady state, we must choose the positive root of α.)

$$\beta = \frac{2A\alpha}{\delta + \rho + 3\alpha} \equiv \beta_m$$

$$\mu = \frac{(A - \beta)}{2\rho}(3\alpha - \delta - \rho) \equiv \mu_m.$$

Note that

$$A - \beta = \frac{A(\delta + \rho + \alpha)}{\delta + \rho + 2\alpha} > 0$$

and $\mu > 0$ if c is sufficiently large. The linear feedback strategy is

$$E = \frac{A(\delta + \rho + \alpha)}{\delta + \rho + 2\alpha} - \alpha S$$

From these, we obtain

$$\dot{S} = \frac{2A(\delta + \rho + \alpha)}{\delta + \rho + 2\alpha} - (2\alpha + \delta)S \tag{1.53}$$

For S to converge to a steady state, it is necessary that $2\alpha + \delta > 0$. It can be verified that this condition is satisfied if and only if the positive root for α is used. Note also that $\alpha > 0$ implies $\beta > 0$. From Eq. (1.53) the steady-state pollution stock under the MPNE with linear feedback strategies is

$$\widehat{S}^M = \frac{2A(\delta + \rho + \alpha)}{(\delta + \rho + 2\alpha)(2\alpha + \delta)} \tag{1.54}$$

Comparing steady states and transient emissions of the transboundary game under OLNE and under MPNE

Comparing the OLNE steady-state pollution stock $\widehat{S}$ (see Eq. (1.16)) with the MPNE steady state $\widehat{S}^M$, we see that the former is smaller than the latter if and only if

$$(\delta + \rho + \alpha)(2c + \delta(\rho + \delta)) > (\delta + \rho)[\delta(\rho + \delta) + 6\alpha^2 + (5\delta + 2\rho)\alpha] \tag{1.55}$$

This relationship always holds, in view of Eq. (1.51). Therefore, $\widehat{S} < \widehat{S}^M$.

Do emissions respond to pollution stock more under the MPNE? Under the MPNE, we have from Eq. (1.53)

$$\frac{\partial \dot{S}^M}{\partial S^M} = -2\alpha - \delta < 0 \tag{1.56}$$

while under the OLNE, we have

$$\frac{\partial \dot{S}^{\text{OL}}}{\partial S^{\text{OL}}} = \lambda_1 = \frac{\rho - \sqrt{(\rho + 2\delta)^2 + 8c}}{2} < 0 \tag{1.57}$$

The absolute value of the RHS of Eq. (1.56) is smaller than that of Eq. (1.57) if and only if

$$\frac{2}{3}\delta + \frac{\rho}{3} + \sqrt{\frac{4}{9}(\rho + 2\delta)^2 + \frac{16}{3}c} < \sqrt{(\rho + 2\delta)^2 + 8c}$$

that is, if and only if

$$\sqrt{\frac{1}{9}(\rho + 2\delta)^2} + \sqrt{\frac{4}{9}(\rho + 2\delta)^2 + \frac{16}{3}c} < \sqrt{(\rho + 2\delta)^2 + 8c}$$

Clearly, this inequality holds for all positive c, ρ, and δ. It follows from this result and from $\widehat{S} < \widehat{S}^M$ that the initial emission under the MPNE is also higher than that under the OLNE.

1.3. Stackelberg Equilibrium

In a two-player game, if one player can make a commitment on what strategy it will use before the other player can choose its strategy, the former is called the Stackelberg leader, and the latter is the follower. In differential games, we make the distinction between open-loop Stackelberg leadership and feedback Stackelberg leadership.

An open-loop Stackelberg leader knows that for any given time path of its control variables, which it announces at the start of the game, the follower will choose a best reply to maximize its payoff. The leader therefore can compute its payoff that would result from each of its feasible announced path, and choose the optimal one. Its best announced path, together with the best reply of the follower, constitute an open-loop stackelberg equilibrium (OLSE).

It turns out that, unlike OLNEs (which are time-consistent), OLSE is generically not time-consistent: if at some time after the game has started the leader is relieved of its commitment to follow its preannounced path, it will typically find it optimal to deviate from that path. (There are exceptions, as we illustrate later.) The intuition behind this time-inconsistency is as follows. If you promise to pay someone a stream of rewards on the condition that it carries out some investment, then once the investment has been sunk, you will have no incentive to keep your promise (given the implicit assumption that there is no loss of reputation, or no cost arising from a loss of reputation).[13] In contrast, in an OLNE, because of the *simultaneous* choice of time paths, no player is trying to influence the action of any other players.

The time-inconsistency of OLSE does not mean that this equilibrium concept is useless. It is a useful equilibrium concept in situations where the leader can credibly precommit, for example by signing a contract that is perfectly enforceable. For example, university teachers in Canada are often required to announce in advance what topics will be covered in the next 13 weeks, in what order, what articles students must read, in what week a midterm exam will be held, what is the percentage of final grade for each assignment, etc. Deviations will be punished (e.g., in some cases promotion can be denied for serious deviations). Thus, a Canadian university teacher is often an open-loop Stackelberg leader, while the students are the followers.

In the next subsection, we consider an example of an OLSE. This is followed by a discussion of feedback stackelberg equilibrium (FBSE).

1.3.1. *Open-loop Stackelberg Equilibrium*

Recall the infinite horizon game of transboundary pollution that we consider in the preceding section. Assume $c_1 = c_2 = c$ and $A_1 = A_2 = A$ for simplicity. Given S_0, consider the OLNE described by the open-loop strategy $E^*(t)$, which is the unique solution of the differential equation

$$\dot{E}^*(t) = \lambda_1 \left[E^*(t) - \widehat{E} \right] \text{ with } E^*(0) = \widehat{E} + \frac{(S_0 - \widehat{S})c}{\lambda_1 - \rho - \delta}, \qquad (1.58)$$

where λ_1, $\widehat{E}$, and $\widehat{S}$ are given by Eq. (1.19), (1.17), and (1.16), and $E^*(0)$ is given by Eq. (1.23). By construction, this strategy will be chosen by

---

[13]The time-inconsistency of open-loop Stackelberg games was pointed out by Simaan and Cruz (1973b, p. 619), and recognized in the macroeconomic literature by Kydland and Prescott (1977). In Kydland and Prescott (1977), the government is the open-loop Stackelberg leader and private individuals are followers. In public economics, the models of optimal redistributive taxation formulated by Chamley (1986) and Judd (1985) are formally open-loop Stackelberg games, and therefore are subject to time inconsistency.

player j if it believes that player i uses the same strategy. Now, suppose the rules of the game change as follows: the players are no longer required to choose their strategies simultaneously. Suppose player 1 is allowed to be the Stackelberg leader and player 2 is the follower. Assume that the leader can credibly commit. Clearly, the leader can achieve the above OLNE outcome, simply by announcing that it will use the open-loop strategy found above, see Eq. (1.58). Therefore, the leader can ensure for itself a payoff at least as great as its OLNE payoff. However, it can do better. It knows that player 2 will not choose the strategy (1.58) if it announces its commitments to a different strategy. In fact the leader knows how the follower would react to any time path $E_1(t)$, by computing its best reply $E_2(t)$ that satisfies the following first-order conditions

$$A - E_2 + \psi_2 = 0$$

$$\dot{S} = E_1 + E_2 - \delta S, \ S(0) = S_0$$

$$\dot{\psi}_2 = cS + (\delta + \rho)\psi_2 \qquad (1.59)$$

as well as the transversality condition

$$\lim_{t \to \infty} e^{-\rho t}\psi_2(t)S(t) = 0 \qquad (1.60)$$

Since E_2 can be expressed as a function of ψ_2, namely $E_2 = A + \psi_2$, the constraints that the leader faces are Eqs. (1.59), (1.60), and

$$\dot{S} = E_1 + A + \psi_2 - \delta S, \ S(0) = S_0 \qquad (1.61)$$

That is, the leader faces two differential equations and a transversality condition.[14] Therefore, the Hamiltonian function for the leader must contain two co-state variables, which we denote by θ and γ, respectively, and which correspond respectively to the "state variables" S and ψ_2. Note that ψ_2 is the follower's shadow price of S, but since the leader faces the differential Eq. (1.59), the variable ψ_2 must be technically treated as a state variable as far as the leader is concerned. The co-state variable γ is then the leader's shadow price of the follower's shadow price. In this model, ψ_2 is the follower's marginal valuation of the pollution stock. This marginal valuation evolves slowly, that is, it must obey the differential Eq. (1.59) and therefore it cannot jump. The more negative is ψ_2, the smaller will be the follower's emissions E_2. The follower's $\psi_2(t)$ depends

---

[14]This solution method was suggested by Simaan and Cruz (1973a).

on its expectation of the whole future time path of emissions by the leader.

Let us formulate and solve the leader's problem. Its Hamiltonian is

$$H = AE_1 - \frac{E_1^2}{2} - \frac{c}{2}S^2 + \theta(E_1 + (A + \psi_2) - \delta S) + \gamma(cS + (\delta + \rho)\psi_2).$$

The necessary conditions are

$$\frac{\partial H}{\partial E_1} = A - E_1 + \theta = 0 \tag{1.62}$$

$$-(\dot{\theta} - \rho\theta) = \frac{\partial H}{\partial S} = -cS - \delta\theta + c\gamma \tag{1.63}$$

$$-(\dot{\gamma} - \rho\gamma) = \frac{\partial H}{\partial \psi_2} = \theta + \gamma(\delta + \rho) \tag{1.64}$$

$$\dot{S} = E_1 + A + \psi_2 - \delta S, S(0) = S_0 \text{ given}$$

or, substituting for E_1

$$\dot{S} = 2A + \theta + \psi_2 - \delta S, \ S(0) = S_0 \text{ given} \tag{1.65}$$

$$\dot{\psi}_2 = cS + (\delta + \rho)\psi_2, \ \psi_2(0) \text{ free} \tag{1.66}$$

In addition, since $\psi_2(0)$ is not exogenously given (it depends on the whole time path of the leader's control variable, E_1), it can be normally chosen by the leader.[15] The leader's optimal choice of $\psi_2(0)$ implies that its initial shadow price for the "state variable" ψ_2 is zero, that is, $\gamma(0) = 0$.[16] Let us write this "transversality condition" as a numbered equation:

$$\gamma(0) = 0 \tag{1.67}$$

---

[15] An important point is that for the Stackelberg leader's problem, one should ask what primitive assumptions of the specific models at hand imply $\psi_2(0)$ can (or cannot) be influenced by the leader. This is a type of "controllability problem", see Xie (1997) and Dockner *et al.* (2000, Chap. 5) for details.

[16] This is technically a transversality condition at the beginning of the time horizon, when the initial state variable is not fixed; see, for example, Léonard and Long (1992, Chap. 7). If you can choose the initial level of a state variable, you should set that variable at the level where its marginal contribution to your objective function is zero. Hence $\gamma(0) = 0$.

In addition, we have the usual transversality conditions at the end of the horizon

$$\lim_{t \to \infty} e^{-\rho t} \theta(t) S(t) = 0$$

$$\lim_{t \to \infty} e^{-\rho t} \gamma(t) = 0$$

as well as the constraint (1.60). These conditions are satisfied as long as θ, γ, S, and ψ_2 converge to finite levels.

Let us focus on the steady state of the system of Eq. (1.63)–(1.66). At the steady state, $\dot{\gamma} = \dot{\theta} = \dot{S} = \psi_2 = 0$, and we have four linear equations:

$$\theta(\delta + \rho) + cS - c\gamma = 0 \tag{1.68}$$

$$\gamma\delta + \theta = 0 \tag{1.69}$$

$$2A + \theta + \psi_2 - \delta S = 0 \tag{1.70}$$

$$cS + (\delta + \rho)\psi_2 = 0. \tag{1.71}$$

Solving these four equations, we get the steady-state pollution stock under the OLSE, which we denote by $\widetilde{S}$:

$$\widetilde{S} = \frac{2A}{\delta + \frac{c}{\rho+\delta} + \frac{c}{\rho+\delta+(c/\delta)}} \tag{1.72}$$

Thus, we have obtained the interesting result that the open-loop Stackelberg game leads to a *higher steady-state pollution stock* than the OLNE $\widehat{S}^a$ in Eq. (1.34). It can be verified that in the OLSE, the leader's level of emissions is greater, and the follower's level is smaller, than in their OLNE counterparts. At the steady state, the leader's shadow price γ of the follower's shadow price is positive:

$$\widetilde{\gamma} = \frac{c\widetilde{S}}{c + \delta(\rho + \delta)} > 0.$$

The leader's steady-state shadow price θ of the pollution stock is

$$\widetilde{\theta} = -\delta\widetilde{\gamma} = -\frac{\delta c\widetilde{S}}{c + \delta(\rho + \delta)} < 0$$

and the follower's steady-state shadow price is

$$\widetilde{\psi}_2 = -\frac{c\widetilde{S}}{\delta + \rho} < 0.$$

Does the steady state have the saddle-point properties? Since there are four differential equations, and two initial values are known (they are $S(0) = S_0$ and $\gamma(0) = 0$ by Eq. (1.67) above), saddle-point stability requires the existence of exactly two negative real roots. This issue was taken up in Long (1992), where an analytical formula for computing the four roots of a more general model is also provided.[17]

Finally, we offer a remark about the lack of time consistency. An OLSE is said to be time inconsistent if, at some time $t_1 > 0$, the leader would want to deviate from the originally announced path once it is no longer required to honor its commitment. Here is a method of showing time inconsistency in our model. The leader's co-state variable $\gamma(t)$ (associated with follower's shadow price ψ_2) is positive at the steady state and therefore is also positive when the steady state is almost reached. If the leader can replan at some time $t_1 > 0$ where $\gamma(t_1) > 0$, it will reset $\gamma(t_1) = 0$, which implies a change in the path for $\theta(t)$, and hence in its emissions. This proves time inconsistency. The intuition is as follows. At the start of the game, the leader would announce that it will emit a lot. This induces the follower to choose a low path of emissions, which is good for the leader. However, once its announced emission plan has "worked", it would no longer want to emit so much. Therefore, at the replanning time t_1, it would choose a lower path of emissions. This reduces the marginal environmental damage to both players. It would be reflected in a less negative value for $\psi_2(t_1)$. The fact that $\gamma(t_1) > 0$ means that at t_1, a new announcement of a lower emission path of the leader would increase its payoffs. In a different context, a similar result on time inconsistency is found in Kemp and Long (1980d), and also Karp (1984), who offers an intuitive explanation of the meaning of the shadow price of the follower's shadow price (p. 80).

1.3.2. *Feedback Stackelberg Equilibrium (FBSE)*

Because of the time inconsistency of OLSE (as noted by Simaan and Cruz, 1973b, p. 619), many authors have attempted to obtain results for leader–follower games using an alternative equilibrium concept: FBSE. This equilibrium requires that the leader must use a feedback strategy: its action (e.g., E_1) at any time t depends on the observed pair $(t, S(t))$, such that

---

[17]The formula originated from Dockner (1985), and has been used by Dockner and Feichtinger (1991) and Kemp *et al.* (1993) to study the existence of limit cycles in optimal control problems and OLSE, respectively.

starting at any (date, state) pair, the continuation of the optimal strategy remains optimal for the leader. The leader must announce its feedback strategy (its decision rule) to the follower, who would believe it only if latter knows that the former would use the same decision rule for all (date, state) pairs. Decision rules that specify action at time t as a function of the observed pair $(t, S(t))$ only are called Markovian decision rules. Unlike OLSE, there is no established general methodology for finding an FBSE.[18]

There is the concept of "stagewise feedback Stackelberg equilibrium", which is most easily explained using a model in discrete time with a finite horizon.[19] At the last period, T, the leader moves first, and the follower responds. Therefore, they both know their equilibrium leader-follower payoffs for period T, as a function of the opening stock x_T. Working backward, in period $T - 1$, again the leader makes the first moves, and the follower responds. This type of dynamic programming approach, based on the assumption that the leader can move first in each period, gives rise to the "stagewise feedback Stackelberg equilibrium".[20] However, in general, the concept of FBSE does not require that in each period the leader takes an action before the follower can move. One can think of the feedback Stackelberg leader as a player who can tell the other player(s): "this is my Markovian decision rule, and you can verify that it is in my interest to use the same decision rule at every point of time." The main difficulty with this "global" Stackelberg leadership is twofold: first, for each Markovian decision rule, we must calculate how the follower will react and second, we must choose from the space of all possible decision rules the one that maximizes the (overall) payoff of the leader. Since this space can be very large, the task of finding a FBSE is formidable.

In some models with very special features, global FBSEs can be found relatively easily. For example, if we can find a strategy for the leader that would achieve the "command and control" outcome (i.e., as if it can control the actions of the other players) then clearly it has no incentive to deviate from it under any circumstances. The optimal pollution tax rule obtained by Benchekroun and Long (1998) for a polluting oligopoly is an instance where the feedback Stackelberg leader (the government) is able to achieve

---

[18]For recent discussions of this issue, see Shimomura and Xie (2008) and Long and Sorger (2009).

[19]The papers by de Zeeuw and van der Ploeg (1991) and Kydland (1975) contain results for stagewise FBSE in discrete time with a finite horizon.

[20]See Başar and Haurie (1984) or Turnovsky *et al.* (1988) for an illustration.

the command and control outcome. Alternatively, one can restrict the space of decision rules that the leader can choose from. Thus, for linear quadratic games, it may seem natural to restrict the leader to decision rules that are linear affine in the state variables. There are several across examples of global Stackelberg leadership and stagewise Stackelberg leadership in the chapters to follow.

Chapter 2

DYNAMIC GAMES IN ENVIRONMENTAL ECONOMICS

Environmental quality (particularly water pollution, airborne particles, soil degradation, and climate change) has increasingly become a major concern. Dynamic models of pollution control appeared in the 1970s (Keeler *et al.*, 1971; Forster, 1972, 1973; Gruver, 1976; Brock, 1977). These models did not consider game-theoretic issues, which often have important consequences. This chapter surveys recent developments in game-theoretic analyses of environmental issues in a dynamic context.

2.1. Models of Transboundary Pollution and Global Warming

The first-generation models of dynamic games of transboundary pollution include those of Kaitala *et al.* (1991, 1992a,b), Long (1992), Ploeg and de Zeeuw (1992), Hoel (1992, 1993), and Dockner and Long (1993). Under fairly general utility and damage cost functions (not restricted to linear quadratic functional form), Long (1992) shows that in the OLNE, countries pollute too much: the OLNE steady-state pollution stock is too high relative to the situation where they cooperate to maximize the sum of their payoffs. Long (1992) also considers the case of OLSE under fairly general assumptions: countries may have different discount rates and different damage cost parameters. An interesting result is that the leader's optimal policy may result in a limit cycle.

Ploeg and de Zeeuw (1992) compute MPNE in linear strategies. Dockner and Long (1993) find that in addition to a unique MPNE where countries use linear feedback strategies, there exists a continuum of MPNE where the countries use non-linear feedback strategies.[1] In particular, there exist pairs of non-linear strategies that approximately achieve the co-operative outcome in the long run. They report analytical equations that

---

[1]The existence of a continuum of Markov-perfect equilibria was first established by Tsutsui and Mino (1990) in a model of duopolistic competition with sticky prices.

implicitly describe these strategies. However, not all these strategies are globally defined. Rowat (2007) finds restrictions on the range of the state variable such that non-linear strategies are well-defined. Zagonari (1998) extends the Dockner–Long model to a pollution control game between environmentally concerned countries and consumption-oriented countries. A dynamic game between asymmetric countries is also considered by Martin *et al.* (1993).

Hoel (1992, 1993) sets up a finite horizon model in discrete time, and shows how equilibria differ when countries play open-loop and feedback strategies. A tax is introduced to achieve efficiency. See also Haurie and Zaccour (1995) in this regard.

List and Mason (2001) consider an asymmetric version of the model of Dockner and Long (1993). They assume there are two regions with different parameters for the regional damage function and production function. They show that the outcome of a game between the two regions may be superior, in terms of social welfare, to that under an imperfect central planner who forces the two regions to have the same emission rate. List and Mason (1999) obtain similar results numerically for a problem where the central planner lacks information about the synergistic effects of two types of pollutants.

Dockner *et al.* (1996) investigate a discrete-time, infinite horizon model of global warming with a catastrophic threshold. The pollution stock, denoted by p_t, is bounded above by $\bar{p}$, interpreted as the critical level beyond which the system will collapse. The damage cost functions become infinite for $p > \bar{p}$. The players are two countries with utility functions that are linear in output (which equals emissions). They prove that if the countries co-operate, the pollution stock will converge, in finite time, to a steady-state level $p^* < \bar{p}$. In contrast, if they do not co-operate, there are two types of MPNE. The first type of equilibrium is called the most rapid approach path (MRAP) equilibrium. The second type of equilibrium, called the "make-the-opponent-indifferent" (MTOI) equilibrium, displays the properties that each country is indifferent among all of its feasible choices. A numerical example is constructed where the MTOI equilibrium generates a chaotic path of pollution.[2] Yanase (2005) generalizes the model of Dockner *et al.* (1996) by allowing the possibility that countries suffer from both flow and stock externalities, where the "flow externalities" may

---

[2]For similar examples, see Dutta and Sunderam (1993b) and Dockner and Nishimura (1999).

include oligopolistic market interactions. He also discusses trigger strategies that ensure co-operative outcomes.

Dutta and Radner (2006, 2009) consider strategic actions in a global warming model set in discrete time and infinite horizon. In their 2006 paper, they consider a world consisting of I countries with exogenous population growth. The damage cost is linear in the stock of pollution. As is well known, when the stock enters the utility function linearly, the control variable (here the rate of emissions), when optimally chosen, will be independent of the stock level.[3] They therefore find that in a Markov-perfect equilibrium, a country's emission is independent of the greenhouse gas (GHG) stock, and dependent only on its own population. Assuming that a country's emission of GHG is related to its energy input by a factor of proportion f_i, they show that if f_i is initially large, a decrease in it will lead to an increase in the equilibrium emission rate.

2.2. Empirical Models of Transboundary Pollution Games

Empirical models are built with the purpose of quantifying the likely impacts of proposed policies on specific countries for a specific period of time. Real-world data are used to calibrate parameters of demand and cost functions. The equilibrium paths of the model are then solved by numerical methods. Early empirical models include those of Kaitala *et al.* (1991, 1992a,b and 1995). These papers deal with transboundary pollution and acid rain games involving a number of countries, including Finland and Russia.

Bernard *et al.* (2008) set up a dynamic game model of trading in pollution permits where the two dominant players are Russia and China.[4] The passive players are other Annex B countries.[5] Russia has a large stock of pollution permits to sell, and China could earn a lot of permits through the "clean development mechanism," one of the three flexibility mechanisms established by the Kyoto Protocol.[6] Russia has a stock of permits $x_1(t)$ that

---

[3]See Dockner *et al.* (2000) for a proof of this result for the class of "linear state games".
[4]A static model of duopoly in the sale of permits is formulated by Loschel and Zhang (2002), where the two players are Eastern Europe and Russia.
[5]Under the Kyoto Protocol, Annex B countries include the OECD countries plus Russia and Central and Eastern European countries.
[6]The other two mechanisms are joint implementation (JI) and international emission trading (IET). Through JI, an Annex B country may implement a project that reduces

are banked at time t. The stock evolves according to the following law of motion

$$x_1(t+1) = x_1(t) - u_1(t) + q_1(t) + h(t)$$

where $u_1(t)$ is the amount of permits it sells in the world market, $q_1(t)$ is its level of emissions abatement, which earns new tradeable permits, and $h(t)$ is its exogenous flow of permits (called *hot air*).[7] The stock $x_2(t)$ of the second dominant player (China) follows a similar law of motion, except that China does not have a hot air flow $h(t)$.

The authors assume that the price of a permit at time t, $p(t)$, is obtained by equating the demand by the rest of the world, $D_R(p(t))$, to the combined supply of permits by Russia and China:

$$D_R(p(t)) = u_1(t) + u_2(t) \tag{2.1}$$

where the function $D_R(.)$ is derived from the competitive equilibrium conditions for the Annex B countries (excluding Russia). This formulation assumes that Annex B countries *do not take expected future prices of permits into account* when they form their current demand. Inverting the market clearing condition (2.1), one obtains

$$p(t) = D_R^{-1}(u_1(t) + u_2(t)) \equiv P(u_1(t) + u_2(t)) \tag{2.2}$$

Both China and Russia know that they can influence the price of permits. So, they behave like duopolists in an exhaustible resource market (here, the resource stocks are $x_1(t)$ and $x_2(t)$). The authors assume that neither China nor Russia care about the damage cost of global warming. Player i maximizes

$$\sum_{t=0}^{T-1} \beta_i^t \left[P(u_1(t) + u_2(t))u_i(t) - c_i(q_i(t)) \right] + \beta_i^T \pi_i x_i(T)$$

where β_i is the discount factor, $c_i(.)$ is the abatement cost function, and π_i is the scrap value per unit of the final stock of permits.[8] This is a difference

---

emissions in another Annex B country. Through IET, an Annex B country may sell emission credits to another Annex B country.

[7] After the disintegration of the Soviet Union, Russia experienced a sharp decline in output, which gives rise to emission credits. These are called hot air because they are available at no cost.

[8] A stochastic variant of the model is that of Haurie and Viguier (2003), where random shocks are independent of the controls.

game (De Zeeuw and van der Ploeg, 1991). The search for an equilibrium is formulated as a non-linear complementary problem, for which efficient algorithms have been developed (Ferris and Munson, 2000).

The authors numerically compute the OLNE. (It would be much more difficult to compute feedback Nash equilibria.)[9] To do the calibration, they use simulation results of a computable general equilibrium model, namely GEMINI-E3 (Bernard and Vielle, 1998, 2000, 2003), and a partial equilibrium model of the world energy system, namely POLES (Criqui, 1996). GEMINI-E3 provides the data to estimate demand for pollution permits by Annex B countries, and marginal abatement costs for Russia. POLES is used to estimate the marginal abatement costs for China.

The Nash equilibrium is then compared with the outcome of an alternative scenario where only Russia behaves strategically.[10] The authors find that duopolistic competition between Russia and China in their permit sales is likely to lower the permit price significantly.

2.3. Carbon Taxes under Bilateral Monopoly

The models mentioned above do not take into account the fact that the additional CO_2 accumulation in the atmosphere comes from the use of fossil fuels, which are extracted from non-renewable stocks (oil fields, coal mines). The dynamics of GHG accumulation is therefore closely linked to the dynamics of resource extraction. To the extent that the world's flow of oil supply is under the control of a cartel, policies to curb GHG emissions (such as carbon taxes) will induce responses by the cartel. This should be taken into account in policy design, as Sinn (2008) recently emphasizes. It is therefore imperative that we develop models to gain insights into the dynamic interactions between tax policies and the pricing or extraction strategies of resource cartels. In what follows, we survey some theoretical attempts in that direction.

The simplest models consider only one-state variable, namely accumulated extraction, and equates it with stock pollution. This may

---

[9]Yang (2003) proposes a method of solving for closed-loop strategies by decomposing them into a sequence of open-loop ones, and implements the algorithm on the RICE model of Nordhaus and Yang (1996), which was formulated as an open-loop game.
[10]Bernard and Vielle (2002) and Bernard et al. (2003) model the dynamic monopoly of Russia in the permit market. Static models of maximizing Russia's rent include those of Bohringer (2001) and Buchner et al. (2002).

be justified on the grounds that the rate of decay of atmospheric CO_2 concentration is very low.

2.3.1. *Nash equilibrium under costless extraction and non-decaying pollution*

Wirl (1994) considers a dynamic game between the government of a fossil-fuel importing country and a monopolist seller of fossil fuels extracted costlessly from a stock of resource $\overline{R}$. It is assumed that the consumption of fossil fuels takes place only in the importing country, and the amount consumed, q, generates two adverse environmental effects. First, there is flow pollution, with damages equal to $(1/2)\eta q^2$ where $\eta \geq 0$ is the flow damage parameter. Second, there is stock pollution, with damages equal to $(1/2)\delta Z^2$ where Z is the state variable that represents the stock of pollution, which is assumed to be the same as accumulated consumption (or accumulated extraction), with $Z(0) = 0$ and $\dot{Z}(t) = q(t)$. Here, it is assumed that the stock of pollution does not decay. The monopolist exporter sets the producer price, p, at each point of time, while the importing country sets a tax rate of τ per unit, such that the consumer price is $p + \tau$. Assume that the demand function of the representative consumer is $q = a - (p + \tau)$. Here $a > 0$ is the choke price, that is, demand is zero if $p + \tau = a$. The consumer's surplus is then

$$\Omega = \frac{1}{2}(a - p - \tau)^2 \quad \text{for } p + \tau \leq a$$

The instantaneous welfare of the importing country at time t is the sum of consumer surplus and the tax revenue, minus the damage costs

$$\begin{aligned} W_M &= \frac{1}{2}(a - p - \tau)^2 + \tau(a - p - \tau) - \frac{1}{2}\eta q^2 - \frac{1}{2}\delta Z^2 \\ &= \frac{1}{2}\left[(a - p)^2 - \tau^2\right] - \frac{1}{2}\eta(a - p - \tau)^2 - \frac{1}{2}\delta Z^2 \end{aligned}$$

The instantaneous welfare of the exporting country, W_X, is assumed to be equal to the seller's profit. Assuming that the cost of extraction is zero, we have

$$W_X = pq = p\left(a - (p + \tau)\right)$$

Let us define instantaneous global welfare by

$$W_G = W_M + W_X$$

Let $\theta \equiv p + \tau$. Then, $q = a - \theta$ and

$$W_G = \frac{1}{2}\left(a^2 - \theta^2\right) - \frac{1}{2}\eta\left(a - \theta\right)^2 - \frac{1}{2}\delta Z^2$$

What would be the price path that a (fictitious) world social planner would choose to maximize the integral of the discounted stream of instantaneous global welfare? Its objective function would be

$$\int_0^\infty e^{-rt}\left[\frac{1}{2}\left(a^2 - \theta^2\right) - \frac{1}{2}\eta\left(a - \theta\right)^2 - \frac{1}{2}\delta Z^2\right]dt$$

subject to

$$\dot{Z} = a - \theta, \quad Z(0) = 0$$

and the constraint that the stock of pollution (cumulative consumption) cannot exceed the stock of exhaustible resource available at time zero:

$$Z(t) \leq \overline{R} \tag{2.3}$$

The Hamiltonian for this problem is

$$H = \frac{1}{2}(a^2 - \theta^2) - \frac{1}{2}\eta\left(a - \theta\right)^2 - \frac{1}{2}\delta Z^2 + \psi\left(a - \theta\right) + \lambda(\overline{R} - Z)$$

where ψ is the co-state variable and $\lambda \geq 0$ is the Lagrange multipler associated with the stock constraint (2.3). In what follows, we assume that the constraint (2.3) is not binding, that is, $Z(t) \to Z_\infty < \overline{R}$. Maximizing H with respect to θ gives the necessary condition

$$-\theta + \eta(a - \theta) - \psi = 0$$

that is, the consumer's price θ is equal to the sum of the marginal flow damage cost, ηq, and the negative of the shadow price of the pollution stock, $-\psi$. It is easy to show that the steady-state stock of pollution under the world social planner is

$$Z_\infty^{so} = \frac{ar}{\delta} \tag{2.4}$$

where the superscript so indicates that this is the socially optimal steady state. Note that this implies that we are assuming that the initial resource stock is large enough, such that

$$\overline{R} \geq \frac{ar}{\delta}$$

At the steady state, the annuity value of marginal damages caused by the stock is just equal to the choke price a:

$$\frac{\delta Z_\infty^{so}}{r} = a$$

Let us turn to a more realistic scenario: there is no world social planner, and the two countries are playing a dynamic game between themselves. Wirl assumes simultaneous moves and seeks a feedback Nash equilibrium. The importing country believes that the exporter (monopolist seller) uses a decision rule $\tilde{\phi}(.)$ that conditions the producer price at time t to the stock size $Z(t)$:

$$p(t) = \tilde{\phi}(Z(t))$$

Recognizing that $p(t) = \tilde{\phi}(Z(t))$, the importing country's instantaneous welfare is

$$W_M(\tau, Z) = \frac{1}{2}\Big[(a - \tilde{\phi}(Z))^2 - \tau^2\Big] - \frac{1}{2}\eta(a - \tilde{\phi}(Z) - \tau)^2 - \frac{1}{2}\delta Z^2$$

Its objective is to maximize its welfare stream, J_M, by choosing its tariff rate $\tau(t)$

$$J_M = \max_{\tau(.)} \int_0^\infty e^{-rt} W_M dt$$

subject to

$$\dot{Z} = a - \tilde{\phi}(Z) - \tau$$

and $Z(0) = 0, Z(t) \leq \overline{R}$. Its HJB equation is

$$rJ_M(Z) = \max_{\tau}\Big\{W_M(\tau, Z) + J'_M(Z)\Big(a - \tilde{\phi}(Z) - \tau\Big)\Big\} \qquad (2.5)$$

The first-order condition is

$$-\tau + \eta(a - \tilde{\phi}(Z) - \tau) - J'_M(Z) = 0 \qquad (2.6)$$

The solution of the importer's HJB equation yields a decision rule $\tau = g(Z)$.

The exporter believes that the importer has a decision rule $\tau = \tilde{g}(Z)$. Its instantaneous welfare (profit) is

$$W_X = pq = p(a - p - \tilde{g}(Z))$$

It then chooses its producer price $p(t)$ to solve

$$J_X = \max_{p(\cdot)} \int_0^\infty e^{-rt} p(a - p - \widetilde{g}(Z)) dt$$

subject to

$$\dot{Z} = a - p - \widetilde{g}(Z)$$

and $Z(0) = 0, Z(t) \leq \overline{R}$. Its HJB equation is

$$rJ_X(Z) = \max_p \{p(a - p - \widetilde{g}(Z)) + J_X'(Z)(a - p - \widetilde{g}(Z))\}$$

The first-order condition is

$$a - 2p - \widetilde{g}(Z) - J_X'(Z) = 0 \tag{2.7}$$

The solution of the exporter's HJB equation yields a decision rule $p = \phi(Z)$. In a Nash equilibrium, expectations are correct, so Eq. (2.6) and (2.7) yield

$$p = \frac{1}{2+\eta}[a + J_M'(Z) - (1+\eta)J_X'(Z)] \tag{2.8}$$

$$\tau = \frac{1}{2+\eta}[a\eta + \eta J_X'(Z) - 2J_M'(Z)] \tag{2.9}$$

It follows that the consumer price is

$$\theta \equiv p + \tau = \frac{1}{2+\eta}[a(1+\eta) - J_M' - J_X'] \tag{2.10}$$

and

$$q = a - (p + \tau) = \frac{1}{2+\eta}(a + J_M' + J_X')$$

$$p + J_X' = \frac{1}{2+\eta}(a + J_M' + J_X')$$

Substituting into the HJB equations, we get a system of two differential equations:

$$rJ_X(Z) = \frac{1}{(2+\eta)^2}(a + J_M' + J_X')^2 \tag{2.11}$$

$$rJ_M(Z) = -\frac{1}{2}\delta Z^2 + \frac{(1+\eta)}{2(2+\eta)^2}(a + J_M' + J_X')^2 \tag{2.12}$$

To solve this system, it is convenient to define the function

$$J(Z) \equiv J_M(Z) + J_X(Z)$$

Then, our problem reduces to solving a single differential equation

$$rJ(Z) = -\frac{1}{2}\delta Z^2 + \gamma\left(a + J'(Z)\right) \tag{2.13}$$

where

$$\gamma \equiv \frac{(3+\eta)}{2(2+\eta)^2}$$

Let us impose the boundary condition that when Z is at the social optimal steady state, that is, $Z = ar/\delta$, extraction is equal to zero and the flow of welfare is simply the damages caused by the stock:

$$rJ\left(\frac{ar}{\delta}\right) = -\frac{1}{2}\delta\left(\frac{ar}{\delta}\right)^2 \tag{2.14}$$

The differential equation (2.13) together with the boundary condition (2.14) has a unique solution:

$$J(Z) = \frac{A}{2}Z^2 + BZ + C$$

where

$$A = \frac{1}{4\gamma}\left[r - \sqrt{r^2 + 2\delta\gamma}\right] < 0$$

$$B = \frac{2a\gamma A}{r - 2\gamma A} < 0$$

$$C = \gamma(a + B)^2 > 0$$

Note that at the steady state, $Z = ar/\delta \equiv Z_\infty^N$, the following equation holds

$$a + J'\left(\frac{ar}{\delta}\right) = a + A\frac{ar}{\delta} + B = 0$$

Thus condition (2.14) is indeed satisfied.

The consumer price under the Nash equilibrium can be obtained from (2.10):

$$\theta = a - \frac{1}{2+\eta}[a + J'(Z)] \tag{2.15}$$

$$= a - \frac{1}{2+\eta}[(a + J'(Z)) - (a + J'(Z_\infty^N))] \tag{2.16}$$

$$= a - \frac{1}{2+\eta}A(Z - Z_\infty^N) \tag{2.17}$$

As the stock of accumulated extraction increases toward the steady-state level, the consumer price rises monotonely to a.

Finally, it can be shown that the equilibrium feedback strategies are, for all $Z \leq ar/\delta$

$$\tau = g(Z) = a + \frac{2A^2 - 2\delta(2+\eta)^2}{r(2+\eta)^3}\left(\frac{ar}{\delta} - Z\right)$$

$$p = \phi(Z) = \frac{\delta(2+\eta)^2 + 2bA^2}{r(2+\eta)^3}\left(\frac{ar}{\delta} - Z\right)$$

Along the equilibrium path, the producer price falls monotonically to zero, and the tax τ rises steadily toward a. It can be shown that the time path of consumer price in the Nash equilibrium is above the one that a world social planner would choose. This is because the seller aims at restricting output in the early phase of the program. Recall Hotelling's famous dictum, "the monopolist is the conservationist's best friend".

Wirl (1994) argues that there exist other Nash equilibria where both players use non-linear strategies, and these equilibria lead to some steady-state $Z_\infty^{**} < ar/\delta$. He also shows that both players would be better off by using linear strategies. One may argue that, in the context of this model, such equilibria are not subgame perfect, in the sense that if both players find themselves at Z_∞^{**}, they would want to move away from it, so that both would gain.

2.3.2. *Stagewise Stackelberg leadership by the seller under costless extraction and non-decaying pollution*

Tahvonen (1996, Sec. 3) considers a leader–follower game using the set-up of Wirl (1994), without the flow externality, that is, $\eta = 0$. The leader is the seller. We should interpret the leadership (as implicitly described in the solution technique used in the model) in the stagewise sense, which we explain below. If the time horizon is finite, and time is discrete, stagewise leadership means that in each period, the leader moves first, and announces the price for that period. The follower then reacts to that price, and chooses the tax for that period. Clearly, this is a well-defined game for the last period, $T - 1$, and each party's equilibrium payoff for period $T - 1$ can be computed as a function of the opening stock Z_{T-1}. Working backward, in period $T - 2$, the leader announces his/her the p_{T-2}, and the follower reacts by choosing τ_{T-2} etc. Stagewise Stackelberg leadership is a well-defined concept, and the solution technique is conceptually simple.

Formulas for calculating stagewise Stackelberg equilibrium for discrete-time games with a finite horizon are given in Kydland (1975) and de Zeeuw and van der Ploeg (1991). Applications include those of Başar *et al.* (1985a), and Turnovsky *et al.* (1988). Extending the stagewise formulation to continuous time horizon involves the feeding of one HJB equation into another, as the following analysis by Tahvonen shows.[11] We demonstrate below that the stagewise Stackelberg equilibrium for this model is identical to the Nash equilibrium found by Wirl. Our demonstration suggests that it might be better to deal with the leadership issue by considering an alternative conception of feedback Stackelberg games, which has been applied by Long and Fujiwara (2010) to the problem of optimal tariff in exhaustible resources.[12]

Following Tahvonen, let us define the stock that remains at time t as $x(t)$. Assume $x(0) = \overline{R}$, and $Z(0) = 0$. Then

$$\overline{R} - x(t) = Z(t) - Z(0) = Z(t)$$

apart from this transformation of variable, we use the same symbols as in the preceding subsection. Further, we continue to assume that $\overline{R} > ar/\delta$. For simplicity, we assume extraction is costless and there is no flow pollution (i.e., $\eta = 0$).

Unlike the formulation of Wirl (1994) where the importing country believes that the seller uses a decision rule $p = \phi(Z) = \phi(\overline{R} - x) \equiv h(x)$ and chooses its tariff τ without knowing (observing) p, in the stagewise Stackelberg formulation the importing country *observes the actual price $p(t)$ before it chooses its tariff rate $\tau(t)$*, and it also knows that $p(t) = h(x(t))$, which is a stationary rule. Therefore, Tahvonen writes the following (time independent) HJB equation of the importing country:

$$rV_M(x) = \max_{\tau} \left\{ \frac{1}{2}(a - p)^2 - \frac{1}{2}\tau^2 - \frac{\delta}{2}(\overline{R} - x)^2 - V_M'(x)\,(a - p - \tau) \right\}$$

$$(2.18)$$

---

[11]Tahvohnen (1996, p. 4) attributes this solution technique to Basar and Haurie (1984).

[12]This alternative conception may be called hierarchical leadership, or leadership in a hierarchical game. It does not assume stagewise leadership (which presupposes that the leader makes a first move in each period, and the follower responds to that move). Instead, the leader in a hierarchical game knows that in response to each of its decision rules, the follower will choose its best-reply decision rule. The leader then chooses among all its possible decision rules the one that maximizes its life-time payoff. This conception is discussed in Dockner *et al.* (2000), and in follow-up papers by Shimomura and Xie (2008) and Long and Sorger (2009).

where $p = h(x)$. Now, *assume that $h(x)$ is linear, that is, $h(x) = \alpha x + \beta$* for some a and β to be determined. Then, it makes sense to conjecture that $V_M(x)$ is quadratic. Let us write

$$V_M(x) = \frac{1}{2} A_M x^2 + B_M x + C_2$$

Notice that the constants A_M, B_M, and C_M should be dependent on the parameters α and β. Maximizing the right-hand side (RHS) of (2.18) gives

$$-\tau + A_M x + B_M = 0$$

Next, the quantity demanded is

$$q = a - p - A_M x - B_M$$

where $p = h(x) = \alpha x + \beta$. It seems that the next logical step is to maximize the leader's integral of the discounted stream of profits with respect to α and β. Instead of doing that directly, Tahvonen follows the method suggested by Basar and Haurie (1984), and writes the following HJB equation for the seller (under costless extraction):

$$r V_X(x) = \max_p \left\{ p \left(a - p - A_M x - B_M \right) - V_X'(x) \left(a - p - A_M x - B_M \right) \right\}$$

Maximizing the RHS with respect to p gives the following first-order condition for the leader:[13]

$$q - p + V_X' = 0 \tag{2.19}$$

Tahvonen conjectures that the leader's value function is also quadratic. Then

$$V_X(x) = \frac{1}{2} A_X x^2 + B_X x + C_X$$

where A_X, B_X, and C_X are constants to be determined, and Eq. (2.19) becomes

$$q - p + A_X x + B_X = 0$$

Thus

$$p = \frac{1}{2} \left[a + (B_X - B_M) + x(A_X - A_M) \right] \tag{2.20}$$

---

[13]This procedure ignores the fact that the leader's choice of the decision rule for setting the wellhead price p would have an effect on A_M and B_M.

and

$$\tau = A_M x + B_M \tag{2.21}$$

Equations (2.20) and (2.21) are exactly the same as Wirl's Nash equilibrium solution, (see Eqs. (2.8) and (2.9)) once we take into account the fact that $V'_M(x)$ and $V'_X(x)$ are mirror images of $J'_M(Z)$ and $J'_X(Z)$. It follows that Tahvonen's procedure of solving for the (stagewise) Stackelberg equilibrium with the exporter as the leader yields a solution that is identical to the Nash equilibrium found by Wirl (1994). It would seem that if one maximizes the leader's life-time welfare with respect to the parameters α and β, one would obtain higher welfare for the leader. This is indeed true, as confirmed by Fujiwara and Long (2010) in a different context.

2.3.3. *Bilateral monopoly with stock-dependent extraction cost and non-decaying pollution*

Rubio and Escriche (2001) consider a variation of Wirl (1995) by assuming that extraction cost increases as the resource stock dwindles. As in Wirl (1995), let $Z(t)$ denote accumulated extraction, and $\dot{Z}(t) = q(t)$. Assume there is no flow pollution. The cost of extracting $q(t)$ is $cZ(t)q(t)$. They consider first the Nash equilibrium. After maximization, the HJB equations yield the following pair of non-linear differential equations[14]

$$rJ_X(Z) = \frac{1}{4}(a - cZ + J'_M + J'_X)^2 \tag{2.22}$$

$$rJ_M(Z) = -\frac{1}{2}\delta Z^2 + \frac{1}{8}(a - cZ + J'_M + J'_X)^2 \tag{2.23}$$

If we set $c = 0$ these equations become identical to Wirl's equations (2.11) and (2.12) with $\eta = 0$, that is, no flow pollution damages. The consumer price, denoted by θ, satisfies

$$\theta = \frac{1}{2}[a + cZ - (J'_M + J'_X)] \tag{2.24}$$

which is the same as (2.15) if $c = \eta = 0$.

Rubio and Escriche (2001) then state that in the Nash equilibrium, the tax τ is only a neutral Pigouvian Tax, in the sense that "it corrects only the inefficiency caused by the stock externality and leaves the cartel with its

---

[14]These are Eqs. (8) and (9) in Rubio and Escriche (2001), with slightly different notation.

monopolistic profit". In proving it, the authors write a Bellman equation for the consumers that they interpret as representing perfectly competitive behavior, which is problematic.[15] In fact that equation, which contains the term $J'_M(Z)q$, implies that consumers are not price takers. Since they know the seller's decision rule for wellhead price, $p = \phi(Z)$, when they choose $q(t)$, they do take into account the fact that their demand $q(t)$ will have an impact on the future level of Z and hence on the price they will pay in the future.

Another point made by Rubio and Escriche (2001) is that (stagewise) Stackelberg equilibrium with the seller as the leader is identical to the Nash equilibrium as described by the three Eqs. (2.22)–(2.24). This result corresponds to the identity between the Nash equilibrium in Wirl (1994) and the (stagewise) Stackelberg equilibrium found by Tahvonen (1996, Section 3).

What happens if the importing country is the leader? Here, Rubio and Escriche (2001) use an approach that indicates stagewise leadership, not hierarchical leadership. Let us outline their analysis.

The exporting country is the follower. At each "period" t, it faces the importer's tax rate $\tau(t)$, which is observed before it chooses the period's wellhead price $p(t)$. Yet, it knows that $\tau(t)$ is determined by the importer's decision rule, $\tau = \widetilde{g}(Z)$. This allows the authors to write a time-independent HJB equation for the exporter:

$$rJ_X(Z) = \max_{p} \left\{ (p - cZ)(a - p - \widetilde{g}(Z)) + J'_X(Z)\left[a - p - g(Z)\right] \right\} \quad (2.25)$$

This gives the exporter's reaction function

$$p = \frac{1}{2}[a - g(Z) + cZ - J'_X(Z)] \equiv p(g, J'_X) \quad (2.26)$$

Plugging this reaction function into the leader's HJB equation, we obtain

$$rJ_M(Z) = \max_{\tau} \left\{ \frac{1}{2}[a - p(g, J'_X)]^2 - \frac{1}{2}\tau^2 \right.$$

$$\left. - \frac{\delta}{2}Z^2 + J'_M(Z)[a - p(g, J'_X) - \tau] \right\} \quad (2.27)$$

---

[15]Liski and Tahvonen (2004) clarify this point.

Maximizing the RHS with respect to τ, bearing in mind that $p(g, J'_X) = p(\tau, J'_X)$, one obtains the first-order condition:[16]

$$-\tau - J'_M(Z) + [a - p]\frac{\partial p}{\partial \tau} - J'_M(Z)\frac{\partial p}{\partial \tau} = 0$$

Hence

$$\tau = \frac{1}{3}[a - cZ - 2J'_M(Z) + J'_X(Z)] \qquad (2.28)$$

Substituting this expression for τ into the follower's reaction function (2.26), one obtains

$$p = \frac{1}{3}[a + 2cZ + J'_M(Z) - 2J'_X(Z)] \qquad (2.29)$$

Inserting Eqs. (2.28) and (2.29) into the RHS of the two HJB equations, one obtains a system of two differential equations

$$rJ_M(Z) = \frac{1}{6}\{a - cZ + J'_X + J'_M\}^2 - \frac{\delta}{2}Z^2$$

$$rJ_X(Z) = \frac{1}{9}\{a - cZ + J'_X + J'_M\}^2$$

To solve this pair of equations, we can apply the technique that we have used to solve the pair (2.11) and (2.12).

The solution reveals that while the long-run stock Z_∞ under stagewise leadership of the importing country coincides with that obtained in the Nash equilibrium, the equilibrium time paths of the stock under the two scenarios are quite different. The initial consumer price and tax are lower in the Nash equilibrium than under the stagewise leadership of the importing country. The life-time payoff of the importing country is higher under its stagewise leadership than under the Nash equilibrium, and the opposite result applies to the exporter's life-time profit.

Liski and Tahvonen (2004) study in more detail the Nash equilibrium of the model of Rubio and Escriche (2001), where the extraction cost rises as the stock of resource dwindles. Their focus is on the interpretation of

---

[16]Note that in doing so, the function g is replaced by the real number τ. It seems that the legitimacy of this procedure needs a rigorous justification. The method of Fujiwara and Long (2010) avoids this difficulty, by not using a HJB equation for the leader. They compute the leader's life-time payoff in terms of the parameters α and β of its decision rule g, and then maximize its life-time payoff with respect to α and β subject to a time-consistency constraint.

the Nash equilibrium carbon tax in terms of a pure Pigouvian motive and a rent-shifting motive. Again, denoting the remaining stock of resource by $x = \overline{R} - Z$, they find that the Nash equilibrium decision rule of the exporter, $p = p_N(x)$, can be an increasing function or a decreasing function. If there are no pollution damages (our $\delta = 0$), then $p_N(x)$ is a decreasing function, that is, the wellhead price will be rising over time (as x falls over time). If the damage costs are very high (δ is sufficiently great), then $p_N(x)$ is an increasing function, that is, the wellhead price will be falling over time. The carbon tax, which consists of a Pigouvian component and a rent-shifting component, decreases (respectively, increases) over time if the damage parameter δ is small (respectively, large).

2.3.4. *Models of carbon taxes with pollution decay*

When the pollution stock has a positive decay rate, there is no longer an one-to-one relationship between the accumulated extraction and the pollution stock. One must then deal with a game involving two-state variables.

The model by Wirl (1995) is an extension of his 1994 paper in two directions: the pollution stock has a constant rate of decay, and the extraction cost increases with cumulative extraction. The game now has two state variables: a stock of cumulative energy production, Z, and a pollution stock, S. Analytical solution does not seem possible with two state variables, hence the author relies on numerical methods. In the special case where the decay rate is zero, the two state variables collapse to one, and analytical solution for a Nash equilibrium becomes possible. It is qualitatively very similar to that found in Wirl (1994). Tahvonen (1996) analyzes a similar game, but under the assumption that the the seller is the stagewise Stackelberg leader.

2.4. International Environmental Agreements

While the motives for international environmental agreements (IEAs) can be studied in a static context, it is important to turn to a dynamic game formulation because stock pollution accumulates over time, and incentives may change when the stock level changes. Another reason for using a dynamic analysis is that because of political institutions, countries may delay joining an agreement, and the effects of delays are of course best studied in an explicitly dynamic framework.

2.4.1. *IEA membership game*

There is an extensive literature on IEAs, beginning with the static models of
Carraro and Siniscalco (1993) and Barrett (1994).[17] Building on the concept
of self-enforcing IEAs introduced by Carraro and Siniscalco (1993) and
Barrett (1994), Rubio and Casino (2005) consider the following two stage
game. In the first stage, countries decide whether to become members of
an IEA. In the second stage, they choose their time paths of emissions over
an infinite horizon.[18] Signatory countries choose their emission paths co-
operatively, while nonsignatories act non-cooperatively. During the emission
stage, countries cannot change their membership status.[19] Using numerical
simulations, Rubio and Casino (2005) find that the only self-enforcing IEA
is a two-country coalition. Germain *et al.* (2003), extending the static
co-operative framework of Chandler and Tulkens (1995), show that it is
possible to devise a dynamic transfer scheme to support the grand coalition.
Petrosjan and Zaccour (2003) also use the co-operative approach to allocate
cost burdens among members of the grand coalition.

Rubio and Ulph (2007) consider a discrete time, infinite horizon
model, where IEA membership can vary over time. They assume that
all countries are identical. After characterizing the fully non-cooperative
equilibrium[20] and the fully co-operative equilibrium, they consider the
case where in each period, there is a two stage game. In stage 1, each
decides whether to be a member of an IEA or not. In stage 2, given
the opening level of the pollution stock, signatories co-operatively choose
their emission level, while each nonsignatory chooses its own emission level
noncooperatively. An IEA is said to be stable (for a given period) if it is both
internally stable and externally stable.[21] At the beginning of each period,
all countries have the same probability of being invited as a signatory. Even
though the payoff in each period is quadratic, in the dynamic membership
framework, the value function is not quadratic. Numerical solutions reveal
three possible patterns of behavior. The most common pattern displays

---

[17]Useful surveys of this literature include Barrett (1997, 2003) and Finus (2001).

[18]Karp and Sacheti (1997) consider a two-period model of an IEA with a stock pollutant
and assess how the incentives to join an IEA are affected by the dynamics of the pollution
stock, and the extent to which the pollution is global.

[19]Many empirical models make the same assumption about commitment to membership,
see, for example, Eyckman (2001).

[20]They focus on the open-loop equilibrium.

[21]Internal stability means that no signatory country would gain to switch to the nonsignatory status; external stability means that no nonsignatory would want to become a
signatory.

a negative relationship between the pollution stock and the number of signatories.

Nkuiya-Mbakop (2009) considers a variant of the model proposed by Rubio and Ulph (2007). He adopts a model in continuous time with an infinite horizon, and treats the length of the period of commitment (to stay with the IEA) as a parameter. Assuming there are n identical countries, he then studies the effect of varying the length θ of the period of commitment on the equilibrium size of the stable coalition and its relationship to the size of the pollution stock. Given the pollution stock size S at the beginning of each period, for any stable IEA of size $n(S)$ in that period, all countries have an equal chance $n(S)/N$ of becoming a signatory. He further shows that there are two critical values of θ, denoted by θ_L and $\theta_H > \theta_L$. If $\theta < \theta_L$, all countries want to be signatories. If $\theta > \theta_H$, the only stable coalition is the two-country coalition. If $\theta \in [\theta_L, \theta_H]$, the greater is θ, the smaller is the size of the stable coalition.

2.4.2. *Effects of delays*

In a world consisting of two identical countries, there is an overwhelming incentive to co-operate, to maximize the sum of their welfare levels. Suppose, however, that for some reasons the two countries cannot enter into an environmental agreement immediately. Using the two-country model of Dockner and Long (1993), Açikgöz and Benchekroun (2005) ask the following question. Suppose the two countries know at time $t = 0$ that at a fixed time T in the future they will, with probability θ, reach an agreement to co-operate from then on, but if that opportunity is missed they would not ever co-operate. What is the effect of that possibility on their emission policies from time zero to time T? The answer depends on whether they use Markov-perfect strategies or open-loop strategies during the phase I, that is during the time interval $[0, T]$.

Each country i seeks to maximize its expected welfare

$$\int_0^T e^{-\rho t} \left[AE_i - \frac{E_i^2}{2} - \frac{c}{2} S^2 \right] dt + e^{-\rho T} [\theta W^C + (1 - \theta) W^{NC}]$$

where W^C and W^{NC} are the respective welfare levels of each country under co-operation and under nonco-operation (the latter depending on whether the OLNE or MPNE will prevail after T given that an agreement is not reached at time T). Here, $c > 0$ is a parameter for pollution damages.

In phase I, if countries use Markov-perfect strategies, their value function will be of the form

$$V_i(S,t) = -\frac{\alpha(t)S^2}{2} - \beta(t)S - \mu(t)$$

where $\alpha(t)$, $\beta(t)$ and $\mu(t)$ are time-dependent parameters. The HJB equation is

$$\rho V_i(S,t) = \max_{E_i} \left\{ AE_i - \frac{E_i^2}{2} - \frac{c}{2}S^2 \right.$$
$$\left. + \left(\frac{\partial V_i}{\partial S}\right)(E_i + E_j(S,t) - \delta S) + \left(\frac{\partial V_i}{\partial t}\right) \right\}$$

where $\rho > 0$ is the discount rate and $\delta > 0$ is the pollution decay rate. The emission strategy must satisfy the first-order condition

$$E_i(S,t) = A + \frac{\partial V_i(S,t)}{\partial S}$$

It follows that $\alpha(t)$ and $\beta(t)$ must satisfy the following differential equations

$$\dot{\alpha} = 3\alpha^2 + \alpha(\rho + 2\delta) - c \tag{2.30}$$

$$\dot{\beta} = \beta(\rho + 3\alpha + \delta) - 2A\alpha \tag{2.31}$$

To solve Eq. (2.30), use the transformation

$$\frac{1}{u(t)} = \alpha(t) - \alpha_m$$

where α_m is the constant solution to (2.30). After solving for $u(t)$, another tranformation gives

$$\alpha(t) = \alpha_m + \frac{2\gamma \exp(2\gamma t)}{K_\alpha - 3\exp(2\gamma t)}$$

where K_α is a constant of integration and

$$\gamma \equiv \sqrt{\left(\delta + \frac{\rho}{2}\right)^2 + 3c}$$

Next, solve for $\beta(t)$, and denote the corresponding constant of integration by K_β. To determine the two constants of integration, we must use the following "smooth pasting" condition at time T, that is,

$$\frac{\partial V_i(S,T)}{\partial S} = \theta \frac{dW^C}{dS} + (1-\theta)\frac{dW^{NC}}{dS}$$

This condition implies that

$$\alpha(T) = \theta\alpha_c + (1-\theta)\alpha_m$$

and

$$\beta(T) = \theta\beta_c + (1-\theta)\beta_m$$

If countries use open-loop strategies in phase I (and assuming that in phase II, if the treaty is not signed, they will also use the open-loop strategies), a similar approach can be used to find the equilibrium of phase I. For the open-loop case, the "smooth pasting" condition is the requirement that the shadow price at T be equal to the weighted average of the two shadow prices under co-operation and under noncooperation after T.

Açikgöz and Benchekroun (2005) find that under the phase I open-loop equilibrium, in anticipation of a possible IEA, countries will pollute more than they would under no anticipation. Further, in the case of Markov-perfect equilibrium in phase I, countries will pollute so much more that the possible welfare gain from co-operation in phase II is almost completely wiped out.

Breton *et al.* (2008) propose a different approach to games of IEAs. Using a discrete-time version of replicator dynamics, they assume that the group of countries that outperform others are copied by a fraction of new entrants to the game. The evolutionary process results in a stable IEA.

2.4.3. *Games under the Kyoto Protocol*

One of the flexible mechanisms encouraged by the Kyoto Protocol is the joint implementation of environmental projects. Breton *et al.* (2005) consider a differential game model of foreign investments in environmental projects. Under the assumption that the damage cost is linear in the pollution stock, they formulate a continuous time, finite-horizon game between two asymmetric countries and compare three scenarios: business-as-usual (BAU), where countries are not committed to any environmental target; autarky, where each country constrains its emissions to achieve a target pollution stock at the end of the horizon; and joint implementations (JIs) where countries can invest in environmental projects abroad. The case of non-linear damage cost is taken up by Breton *et al.* (2006) in a linear quadratic model. They find that if joint implementation is allowed, both countries will invest in environmental projects at home, and at least one will invest abroad. The identity of the latter can change at most once during the planning horizon. For other contributions to the analysis of game theoretic aspects of international co-operations, see Kaitala *et al.* (1995), Kaitala and Pohjola (1995), and de Zeeuw (1998).

2.5. Taxation Issues

2.5.1. *Efficiency-inducing tax on emissions: monopoly or oligopoly as follower and government as leader*

Benchekroun and Long (1998) ask the question: assuming that emissions by firms in an oligopoly (or, as a special case, a monopoly) contribute to the accumulation of a stock of pollution that causes damages, can a benevolent government impose a tax on emissions such that the outcome at each point in time is the same as under the first-best scenario?[22] In the case of symmetric firms, they find that the answer is in the affirmative. They show the existence of a feedback rule for a per unit tax on emissions such that the oligopoly will achieve the social optimum.[23]

Formally, the model of Benchekroun and Long (1998) is a feedback Stackelberg game, with a leader (the government) and n followers (the firms). The followers take the tax rule as given, and play a dynamic Cournot game among themselves. Thus, for any given tax rule, the firms seek output strategies (which describe how much each firm will produce at any time as a function of the observed stock of pollution at that time) that constitute an MPNE (among the firms). The government, knowing how the MPNE depends on the tax rule, chooses the best tax rule (the one that maximizes social welfare). In this model, the best tax rule actually induces firms to produce as if they were operating in a control-and-command economy under

---

[22]In the first best scenario, a hypothetical central planner has the power to control and command firms to collectively maximize social welfare.

[23]Batabyal (1996a,b) considers a leader–follower model where the leader is the regulator and the follower is an industry that pollutes, either under perfect competition or under monopoly. The papers do not specify how the industry's profit may appear as part of the regulator's conception of social welfare. To obtain time-consistency, use is made of the method suggested by Karp (1984). For an exposition of this method, see Chap. 5 of this book.

a benevolent dictator. In a linear quadratic model, the social optimum emission rule is linear in the pollution stock. The optimum can be found by choosing the correct slope and the correct intercept of for the emission rule. In an MPNE among firms, the feedback outcome can also be represented by a linear emission rule. The social planner can influence the slope and the intercept of this rule by choosing a per unit tax rule that is linear in the pollution stock. Thus to achieve the social optimum, one needs to find only two correct numbers. Therefore, while in general it is difficult to find an FBSE, in this relatively simple model, the problem can be easily solved.

In a related paper, Benchekroun and Long (2002) show that there is a multiplicity of efficiency-inducing tax rules. All these rules achieve allocative efficiency; only the after-tax profits are different.

Let us illustrate their argument using a simple model of pollution. Let S be the stock of pollution, and $D(S)$ be the damage function. Emissions are proportional to output, which we denote by Q. Let $P(Q)$ be the inverse demand function for the good, and $U(Q)$ be the area under the inverse demand function, which may be called the gross utility of consuming Q. For simplicity, assume the production cost is zero. Suppose the social planner wants to maximize

$$\int_0^\infty e^{-\rho t} \left[U(Q) - D(S) \right] dt$$

subject to $\dot{S} = Q - \delta S$. Let ψ be the co-state variable. The necessary conditions are

$$P(Q) + \psi = 0$$
$$\dot{\psi} = (\rho + \delta)\psi + D'(S)$$

and the transversality condition is

$$\lim_{t \to \infty} e^{-\rho t} \psi(t) S(t) = 0$$

Assume there exists a unique value $S_\infty > 0$ which satisfies the steady-state condition

$$P(\delta S_\infty)(\rho + \delta) = D'(S_\infty) \tag{2.32}$$

The Euler's equation is

$$P'(Q)\dot{Q} = (\rho + \delta)P(Q) - D'(S)$$

Hence

$$\frac{dQ}{dS} = \frac{\dot{Q}}{\dot{S}} = \frac{(\rho + \delta)P(Q) - D'(S)}{(Q - \delta S)\,P'(Q)} \tag{2.33}$$

This differential equation, together with the boundary condition (2.32) determines the central planner's optimal policy function $Q^c(S)$, where the superscript denotes the outcome achieved under the central planner. Another method of obtaining this differential equation is via the use of the HJB equation

$$\rho V(S) = \max_{Q}[U(Q) - D(S) + V'(S)\,(Q - \delta S)]$$

The first-order condition gives Q as a function of S:

$$P(Q) + V'(S) = 0$$

Differentiation with respect to S yields

$$P'(Q)Q'(S) + V''(S) = 0$$

On the other hand, differentiating the HJB equation, and making use of the envelope theorem, we obtain

$$\rho V'(S) = -D'(S) - \delta V'(S) + (Q - \delta S)\,V''(S)$$

Hence

$$-(\rho + \delta)P(Q) = -D'(S) - (Q - \delta S)\,P'(Q)Q'(S) \tag{2.34}$$

which is identical to Eq. (2.33).

Now suppose the output is produced by a monopolist facing a per unit tax $\theta(S)$. Its Hamiltonian function is

$$H = P(Q)Q - \theta(S)Q + \psi_m(Q - \delta S)$$

The necessary conditions are

$$P'(Q)Q + P(Q) - \theta(S) + \psi_m = 0 \tag{2.35}$$

$$\dot{\psi}_m = (\rho + \delta)\psi_m + \theta'(S)Q \tag{2.36}$$

and the transversality condition is

$$\lim_{t \to \infty} e^{-\rho t}\psi(t)S(t) = 0$$

Denote the steady-state stock under monopoly by S_∞^m. Then

$$[P'(\delta S_\infty^m)\delta S_\infty^m + P(\delta S_\infty^m) - \theta(S_\infty^m)] (\rho + \delta) = \theta'(S_\infty^m)\delta S_\infty^m \qquad (2.37)$$

Differentiating Eq. (2.35) with respect to t gives

$$[P''(Q)Q + 2P'(Q)]\dot{Q} - \theta'(S)\dot{S} = -\dot{\psi}_m = -(\rho + \delta)\psi_m - \theta'(S)Q$$
$$= [P'(Q)Q + P(Q) - \theta(S)](\rho + \delta) - \theta'(S)Q$$

Therefore

$$[P''(Q)Q + 2P'(Q)]\dot{Q} = [P'(Q)Q + P(Q) - \theta(S)](\rho + \delta) - \delta S\theta'(S)$$

and finally, dividing by $\dot{S}$

$$\frac{\dot{Q}}{\dot{S}} = \frac{dQ}{dS} = \frac{[P'(Q)Q + P(Q) - \theta(S)](\rho + \delta) - \delta S\theta'(S)}{[P''(Q)Q + 2P'(Q)](Q - \delta S)} \qquad (2.38)$$

This differential equation, together with the boundary condition (2.37), determines the monopolist's optimal control in feedback form, which we denote by $Q^m(S)$, given the tax function $\theta(S)$.

Now, by inspecting Eqs. (2.37) and (2.38), we see that if we replace $\theta(S)$ by $\theta(S) + g_K(S)$ where $g_K(S)$ satisfies $(\rho + \delta)g_K(S) + \delta S g_K'(S) = 0$, then the original solution $Q^m(S)$ remains the monopolist's optimal feedback control rule. Note that $(\rho + \delta)g_K(S) + \delta S g_K'(S) = 0$ implies $g_K(S) = Ke^{-(\rho+\delta)/\delta}$.

Finally, the central planner can choose $\theta(S)$ such that $Q^m(S)$ coincides with $Q^c(S)$, by choosing a function $\theta(S)$ that satisfies the equation

$$[P''(Q^c)Q^c + 2P'(Q^c)](Q^c - \delta S)\frac{dQ^c}{dS} - [P'(Q^c)Q^c + P(Q^c)](\rho + \delta)$$
$$= -\theta(S)(\rho + \delta) - \delta S\theta'(S) \qquad (2.39)$$

and the boundary condition (2.37). As an example, consider the linear demand case, $P = A - Q$. Assume $D(S) = \gamma S^2/2$. The central planner's optimal steady state is

$$S_\infty = \frac{A(\rho + \delta)}{\gamma + \delta(\rho + \delta)}$$

and the optimal emission rule, in feedback form, is

$$Q^c(S) = \alpha - \beta S \qquad (2.40)$$

where we can use Eq. (2.34) to determine β and α by the method of undetermined coefficients. Thus, from Eqs. (2.34) and (2.40), we get

$$\beta(\alpha - \beta S - \delta S) - (A - \alpha + \beta S)(\rho + \delta) = -\gamma S \qquad (2.41)$$

This equation must hold for all S. This implies that

$$\beta = \frac{-\rho - 2\delta + \sqrt{(\rho + 2\delta)^2 + 4\gamma}}{2} > 0$$

$$\alpha = \frac{(\rho + \delta)A}{\beta + (\rho + \delta)}$$

where the positive root of β has been chosen for convergence. The feedback rule (2.40), with $\beta > 0$, makes sense: the economy should reduce its emissions as the stock of pollution accumulates.

Now, to make the monopolist achieve the social optimum, we use Eq. (2.39) to get

$$2\beta(\alpha - \beta S - \delta S) - 2(A - \alpha + \beta S)(\rho + \delta) + A(\rho + \delta)$$
$$= -\theta(S)(\rho + \delta) - \delta S \theta'(S)$$

Hence, making use of (2.41)

$$-2\gamma S + A(\rho + \delta) = -\theta(S)(\rho + \delta) - \delta S \theta'(S) \qquad (2.42)$$

Clearly, a linear tax rule of the form $\theta(S) = \lambda + \mu S$ would satisfy Eq. (2.41) if and only if

$$\lambda = -A$$

$$\mu = \frac{2\gamma}{\rho + 2\delta}$$

But we can add to this tax function another term, $KS^{-(\rho+\delta)/\delta}$ and still satisfy Eq. (2.39).[24]

---

[24]This example, where the monopolist maximizes the value of the integral of discounted profit over an infinite horizon, taking as given a feedback tax rule announced by the government, and the latter chooses a tax rule to maximize the integral of discounted social welfare, is another illustration of how an FBSE may be found.

It is important to note that in the linear quadratic model, adding $KS^{-(\rho+\delta)/\delta}$ to the tax function will result in a different value function for the monopolist's HJB equation

$$rV(S) = \max_Q \left[P(Q)Q - (\theta(S) + KS^{-(\rho+\delta)/\delta})Q + \frac{dV(S)}{dS}(Q - \delta Q) \right]$$

If $K > 0$, the greater K is, the lower the monopolist's life-time profit. It can be verified that the new HJB equation implies the same decision rule $Q^m(S) = \alpha - \beta S$. The argument can be generalized to the non-linear-quadratic case.

2.5.2. *Prices versus quantities*

When it comes to regulating firms when the regulator does not have full information, are price instruments better than quantity instruments? Weitzman (1974) has considered this issue in a static context. A recent contribution on this issue, in a context where firms behave strategically, is Moledina *et al.* (2003). The authors first consider a two-period model, where firms try to manipulate the regulator's beliefs about their abatement costs. The regulator in this model is naïve: he does not know that firms behave strategically. In the case of an emission tax, the firms will over-abate in period one to induce the regulator to set a lower tax in period two. It should be noted that Moledina *et al.* (2003) are not dealing with a mechanism design problem.

2.5.3. *Tax adjustment rule to achieve a long-run pollution target*

Karp and Livernois (1994) consider a pollution tax adjustment rule that would achieve a long-run pollution target when the regulator does not have full information about abatement costs. In their model, pollution is a flow: it does not accumulate. It is local, not global. There are n firms in the local region. They have no market power in the goods market, but they behave strategically because they know that they can influence the future pollution tax rate. The regulator is not maximizing any objective function. He/she just wants to achieve a long-run pollution target X^*. Let $x_i(t)$ be firm i's emissions at time t. Aggregate emissions are denoted by $X(t) = \sum_i x_i(t)$. Let $s(t)$ be the tax per unit of emissions. Firm i's profit (net of tax) is $\pi_i(t) = R_i(x_i(t)) - s(t)x_i(t)$. The regulator uses the following tax adjustment rule

$$\dot{s}(t) = \begin{cases} \alpha(X(t) - X^*) & \text{if } s(t) > 0 \text{ or } s(t) = 0 \quad \text{and} \quad X(t) \geq X^* \\ 0 & \text{if } s(t) = 0 \quad \text{and} \quad X(t) < X^* \end{cases}$$

where α is a positive constant, exogenously given. Each firm, fully aware of the tax adjustment rule, chooses the time path of $x_i(t)$ to maximize the integral of discounted profit, taking as given the strategies of other firms. They are playing a differential game among themselves.

Karp and Livernois show that if firms are identical and play open-loop strategies, in the steady state the per unit tax, s_∞, is less than the full-information tax. If firms are not identical, then under the OLNE the steady-state allocation of emissions among firms is not efficient. If firms use feedback strategies, there is a continuum of equilibria even in the case of identical firms. If firms are not identical, under certain conditions there exist candidate Markov-perfect equilibria that allocate emissions efficiently at the steady state. However, under the proposed tax adjustment rule, efficiency is unlikely outside the steady state.

2.5.4. *Non-point-source pollution taxes*

The standard Pigouvian tax for pollution discharges relies on the assumption that the authority can observe the rate of discharge of each polluter. When such observations are not available or too costly, the authority could rely on a tax scheme whereby a firm's pollution tax liability depends only on aggregate emission of the whole industry. In this case, is it possible to design a tax scheme that would induce firms to collectively achieve the socially optimal rate of abatement of pollution? An analysis of this problem in a static setting was provided by Segerson (1988) who, drawing on Holmstrom's (1982) analysis of moral hazard in teams, proposed that firms be charged a unit tax that depends on the overall level of pollution, the so-called "ambient tax". For example, a monitoring station can only measure the aggregate pollution of a group of farmers in the same watershed. Xepapadeas (1992) extends the analysis to a dynamic framework, but under the assumption that the taxing authority's aim is to make firms achieve *only the long-run* optimal pollution stock. Whether there is a tax scheme that would induce firms to achieve the efficient time path of pollution *both* in the short run and in the long run remains an open question. Karp (2005) studies the tax revenue implications when the regulator adjusts the tax rate linearly with respect to the deviation of actual pollution from the socially efficient level. He focuses on open-loop equilibrium and flow pollution.

2.6. Environmental Games with Coupled Constraints

In some resource and environmental problems, the set of actions that a player can take may be constrained by the action of other players. For example, consider a static water use problem: since there is only so much water in an open-access well, the total use by all players cannot exceed the total amount of water available. Similarly, if the parties to an international treaty agree on a fixed ceiling on aggregate emissions but there is no mechanism for allocation of emissions rights among them, how would the parties choose individual emission levels while still wishing to respect the ceiling? Constraints of this type are often referred to as coupled constraints (see, e.g., Haurie, 1995; Haurie and Zaccour, 1995; Carlson and Haurie, 2000; Tidball and Zaccour, 2005; Krawczyk, 2005; Bahn and Haurie, 2008). Coupled constraints have important implications for the set of possible equilibria. In particular, there may exist a continuum of Nash equilibria. Rosen (1965) proposes the concept of normalized Nash equilibrium to pick out a unique solution. Krawczyk (2005) regrets that references to this concept are rare in the "main-stream economic" literature, and defends the usefulness of this equilibrium concept.[25] To understand these issues, it is best to start with a simple static game.

2.6.1. *A static model of global pollution with and without a coupled constraint*

Consider a world consisting of two countries, called Home and Foreign respectively. For the moment, we abstract from dynamic considerations by assuming that there is only one period.

There is a homogeneous consumption good. Let x and X denote respectively the home and foreign outputs of this good. One unit of output generates one unit of emission of CO_2. Global pollution is therefore $\pi = x + X$. The utililty function of the home country is

$$u = ax - \frac{b}{2}x^2 - \frac{s}{2}\pi^2$$

---

[25]Krawczyk (2005) gives a clear explanation of the concept of coupled constraint Nash equilibrium, and presents an example of a river basin pollution game, which he solves numerically.

where $a > 0$, $b > 0$ and $s > 0$ are parameters. We assume that a is large relative to b and s. Similarly, the utility function of the foreign country is

$$U = AX - \frac{B}{2}X^2 - \frac{S}{2}\pi^2$$

(Thus we use lower case roman letters for variables and parameters relating to the home country, and upper case roman letters for the foreign country; Greek letters represent variables common to both countries.)

Substituting $x + X$ for π, we get the payoff functions of the two countries:

$$u = f(x, X) = ax - \frac{b}{2}x^2 - \frac{s}{2}(x + X)^2 \tag{2.43}$$

$$U = F(x, X) = AX - \frac{B}{2}X^2 - \frac{S}{2}(x + X)^2 \tag{2.44}$$

A Nash equilibrium (of the unconstrained pollution problem) is a pair (x^*, X^*) such that

$$f(x^*, X^*) \geq f(x, X^*) \quad \text{for all } x$$

$$F(x^*, X^*) \geq F(x^*, X) \quad \text{for all } X$$

The FOCs that characterize the Nash equilibrium are

$$\frac{\partial f}{\partial x} = a - bx - s(x + X) = 0$$

$$\frac{\partial F}{\partial X} = A - BX - S(x + X) = 0$$

(notice that the foreign parameters A, B, and S do not appear in the home country's first-order condition, and vice versa). Solving, we get

$$x^* = \frac{aB + (a - A)S}{bB + bS + sB}$$

$$X^* = \frac{Ab + (A - a)s}{bB + bS + sB}$$

The total pollution is

$$\pi^* = x^* + X^* = \frac{aB + (a - A)(S - s)}{bB + bS + sB}$$

The Nash equilibrium of this game (which does not have an aggregate constraint on emissions) is unique. See, for example, Theorem 2 of Rosen (1965).

Numerical Example 1: Let $a = A = b = B = s = S = 1$. Then

$$x^* = X^* = \frac{1}{3}$$

$$\pi^* = x^* + X^* = \frac{2}{3}$$

Let us compare the Nash equilibrium pollution π^* with what a world planner would want. We assume that the world planner maximizes a weighted sum of the payoffs of the two countries, with the weights being $r \geq 0$ and $R \geq 0$, respectively. We assume that r and R are exogenous. The world planner chooses x and X to maximize

$$\phi(x, X) = rf(x, X) + RF(x, X)$$

where the functions f and F are defined by (2.43) and (2.44). The FOCs are

$$\frac{\partial \phi}{\partial x} = r[a - bx - s(x + X)] - RS(x + X) = 0$$

$$\frac{\partial \phi}{\partial X} = R[A - BX - S(x + X)] - rs(x + X) = 0$$

Consider the simplest case where $R = r = 1$. The socially optimal outputs are

$$x^0 = \frac{aB + (s + S)(a - A)}{Bb + (b + B)(s + S)}$$

$$y^0 = \frac{Ab + (s + S)(A - a)}{Bb + (b + B)(s + S)}$$

and the total emission is

$$\pi^0 = x^0 + X^0 = \frac{aB + Ab}{Bb + (b + B)(s + S)} = \frac{aB + Ab}{(bB + bS + sB) + (BS + bs)}$$

Thus $\pi^0 < \pi^*$, that is, the Nash equilibrium total emission is higher than the socially optimal level.

Numerical Example 2: Use the parameters in Numerical Example 1. Then

$$x^0 = X^0 = \frac{1}{5}$$

$$\pi^0 = \frac{2}{5}$$

Introducing a coupled constraint

Suppose the social planner cannot control emissions of individual countries, but can impose an aggregate constraint on total emissions:

$$x + X \leq \overline{\pi}$$

where $\overline{\pi}$ is an upper bound on total emissions. Such a constraint is called a "coupled constraint". Suppose that the two countries must satisfy this coupled constraint. What is the Nash equilibrium when this constraint must be respected? In what follows, we assume that $\overline{\pi} < \pi^*$.

Nash equilibria with a coupled constraint: non-uniqueness

The home country must choose x to maximize $f(x, X)$ subject to $x + X \leq \overline{\pi}$. Let λ_1 be its Lagrange multiplier associated with the coupled constraint. The Lagrangian is

$$\mathcal{L}_1 = f(x, X) + \lambda_1 (\overline{\pi} - x - X)$$

The FOCs are

$$\frac{\partial \mathcal{L}_1}{\partial x} = a - bx - s(x + X) - \lambda_1 = 0$$

$$\frac{\partial \mathcal{L}_1}{\partial \lambda_1} = \overline{\pi} - x - X \geq 0, \, \lambda_1 \geq 0, \lambda_1 (\overline{\pi} - x - X) = 0$$

Similarly, for the foreign country

$$\mathcal{L}_2 = F(x, X) + \lambda_2 (\overline{\pi} - x - X)$$

$$\frac{\partial \mathcal{L}_2}{\partial X} = A - BX - S(x + X) - \lambda_2 = 0$$

$$\frac{\partial \mathcal{L}_2}{\partial \lambda_2} = \overline{\pi} - x - X \geq 0, \, \lambda_2 \geq 0, \lambda_2 (\overline{\pi} - x - X) = 0$$

We will denote the Nash equilibrium values of this coupled-constraint game with the superscript $\#$. We want to compare $x^{\#}$ with x^*.

It turns out that there can be many Nash equilibria for this coupled-constraint game. Let us consider the case where the coupled constraint is binding. Then, we have three equations with four unknowns:

$$a - bx - s(x + X) - \lambda_1 = 0 \quad \text{where } \lambda_1 \geq 0$$
$$A - BX - S(x + X) - \lambda_2 = 0 \quad \text{where } \lambda_2 \geq 0$$
$$\overline{\pi} - x - X = 0$$

Numerical Example 3: Use the parameters in Numerical Example 1. Assume $\overline{\pi} = 2/5$. One Nash equilibrium of this coupled constraint game is

$$x^\# = X^\# = \frac{1}{5}, \quad \pi^\# = \frac{2}{5}$$

It is associated with $\lambda_1 = \lambda_2 = 0.4$.

Here is another Nash equilibrium

$$x^{\#\#} = \frac{9}{50}, \quad X^{\#\#} = \frac{11}{50}, \quad \pi^{\#\#} = \frac{2}{5}$$

It is associated with $\lambda_1 = 31/50$ and $\lambda_2 = 29/50$. In fact, there is a continuum of equilibria. Rosen (1965) gave other examples of non-uniqueness of this type.

Rosen's normalized equilibrium

Rosen showed that each Nash equilibrium (in a continuum of Nash equilibria) corresponds to a "normalized equilibrium", which he defined as follows.

Take an arbitrary weight vector $\rho \equiv (\rho_1, \rho_2) \geq (0,0)$. Define the function

$$\theta(\overline{x}, \overline{X}, x, X, \rho) \equiv \rho_1 f(x, \overline{X}) + \rho_2 F(\overline{x}, X)$$

Consider the following problem: Given $(\overline{x}, \overline{X}, \overline{\pi}, \rho)$, find (x, X) to maximize $\theta(\overline{x}, \overline{X}, x, X, \rho)$ subject to $x + X \leq \overline{\pi}$. Assume that $f(x, \overline{X})$ is strictly concave in x, and that $F(\overline{x}, X)$ is strictly concave in X. For *fixed* $(\overline{x}, \overline{X}, \overline{\pi}, \rho)$, this problem yields a *unique* solution

$$\widehat{x} = \widehat{x}(\overline{x}, \overline{X}; \overline{\pi}, \rho) \quad \text{and} \quad \widehat{X} = \widehat{X}(\overline{x}, \overline{X}; \overline{\pi}, \rho) \qquad (2.45)$$

A normalized equilibrium is defined as a fixed point of the (hat) mapping described by (2.45), that is, it is a point $(\overline{x}^{**}, \overline{X}^{**})$ such that

$$\widehat{x}(\overline{x}^{**}, \overline{X}^{**}; \overline{\pi}, \rho) = \overline{x}^{**} \quad \text{and} \quad \widehat{X}(\overline{x}^{**}, \overline{X}^{**}; \overline{\pi}, \rho) = \overline{X}^{**}$$

Theorem 3 in Rosen (1965) states that if each player's payoff function is concave in his choice variable then, for any fixed $\rho \equiv (\rho_1, \rho_2) \geq (0,0)$, there exists a normalized equilibrium.

How about uniqueness of normalized equilibrium for a fixed $\rho \equiv (\rho_1, \rho_2) \geq (0,0)$? For this purpose, define the function

$$\sigma(x, X, \rho) \equiv \rho_1 f(x, X) + \rho_2 F(x, X)$$

(notice that there is no upper-bar variables in this definition). The function $\sigma(x, X, \rho)$ is said to be diagonally strictly concave in (x, X) if and only if for any (x', X') and (x, X) the following inequality holds

$$\begin{bmatrix} x - x' & X - X' \end{bmatrix} \begin{bmatrix} \rho_1 f_x(x', X') \\ \rho_2 F_X(x', X') \end{bmatrix} + \begin{bmatrix} x' - x & X' - X \end{bmatrix} \begin{bmatrix} \rho_1 f_x(x, X) \\ \rho_2 F_X(x, X) \end{bmatrix} > 0$$

In his Theorem 4, Rosen (1965) showed that if $\sigma(x, X, \rho)$ is diagonally strictly concave then for any fixed ρ, the normalized equilibrium is unique.

Rosen (Theorem 5) showed that the payoff to player i in normalized equilibrium is in some sense an increasing function of the weight ρ_i.

How do normalized equilibria compare with Pareto efficient allocations?

It is important to note that normalized equilibria (for a given aggregate emission constraint $\overline{\pi} > 0$) may not be Pareto efficient, because in the definition of the normalized equilibrium, the maximization does not fully take into account the externalities.

Consider the following problem for the social planner: for given fixed weights $\gamma_1 > 0$ and $\gamma_2 > 0$, choose x and X to maximize the weighted social welfare function

$$\Omega \equiv \gamma_1 f(x, X) + \gamma_2 F(x, X)$$

The first-order conditions are

$$\gamma_1 f_x(x, X) + \gamma_2 F_x(x, X) = 0$$
$$\gamma_1 f_X(x, X) + \gamma_2 F_X(x, X) = 0$$

These equations determine optimal emissions x^0, X^0, and $\pi^0 = x^0 + X^0$.
Then

$$f_x(x^0, X^0) = -\frac{\gamma_2}{\gamma_1} F_x(x^0, X^0)$$

and

$$F_X(x^0, X^0) = -\frac{\gamma_1}{\gamma_2} f_X(x^0, X^0)$$

Suppose we set the coupled constraint $x + X \leq \pi^0$. Is there a normalized (Rosen) equilibrium that achieves the social optimum (x^0, X^0)?

The first-order conditions for a Rosen normalized equilibrium with weights ρ_1 and ρ_2 are

$$f_x(x, X) = \frac{\lambda}{\rho_1}$$

$$F_X(x, X) = \frac{\lambda}{\rho_2}$$

Now we can state the following result:

Theorem (Cirano, 2009)[26]

If we set ρ_1 and ρ_2 such that

$$\rho_1 = \frac{-\gamma_1}{\gamma_2 F_x(x^0, X^0)} \quad \text{and} \quad \rho_2 = \frac{-\gamma_2}{\gamma_1 f_x(x^0, X^0)} \qquad (2.46)$$

and impose the coupled constraint $x + X \leq \pi^0$, then the Rosen normalized equilibrium with weights (ρ_1, ρ_2) mimicks the solution of the world planner welfare function with weights (γ_1, γ_2). In general the condition (2.46) implies $\gamma_1/\gamma_2 \neq \rho_1/\rho_2$, except for the symmetric case where f and F satisfy

$$f(y, z) = F(z, y) \quad \text{for all } z, y.$$

This theorem can be generalized to the case of many players.

2.6.2. *Models of environmental games with coupled constraints*

As the above exposition of Rosen's normalized equilibrium indicates, a major problem with the concept of normalized equilibrium is that it is not clear, in actual applications, how the weights are determined. Who has the power to fix the weights? Notwithstanding this concern, there are static and dynamic games with coupled constraints that use this concept, and interesting results do emerge. Examples in environmental economics

---

[26]This theorem was discovered by the discussant of a paper presented at a workshop on environmental economics organized by Cirano in Montreal, in April 2009.

include those of Haurie and Krawczyk (1997), Krawczyk (2005), Tidball and Zaccour (2005), and Bahn and Haurie (2008).

Krawczyk (2005) proposes a static river basin pollution game, which presents a non-point source pollution problem, and extends it to a dynamic framework. An open-loop equilibrium is computed.

Bahn and Haurie (2008), building on Haurie *et al.* (2006) and Bahn *et al.* (2008), propose a discrete-time, finite-horozon model of negotiation of a self-enforcing agreement on global GHG emissions. This is called the game of sharing an emission budget. The model considers different groups of countries at different levels of development. The authors assume that an international agreement sets a global environmental constraint, and that parties to the agreement noncooperatively choose their own emission levels while still wishing to satisfy the aggregate constraint.[27] The paper uses the concept of normalized equilibrium proposed by Rosen (1965). Bahn and Haurie (2008) consider m players (called regions), each being a group of countries. Each region is described by a growth model consisting of two sectors, one called the carbon economy, the other called the clean economy, where a much lower level of emissions is necessary in the (more expensive) production process. The open-loop information structure is assumed. A strategy for region j consists of (i) a sequence of investment in capital stocks of its two sectors, (ii) a sequence of labor allocations, and (iii) a sequence of caps on its own emssions. To calculate the normalized equilibrium, each region is given an exogenous weight. Using Rosen's approach, a fixed point of the best-reply mapping under these weights is computed for a two-region model. Region 1 consists of all the developed countries, while region 2 is the set of developing countries. The authors use mostly parameter values from the DICE model of Nordhaus (1994). Giving the two regions the same weight, the authors obtain the Rosen equilibrium and compare it with the Pareto optimum (i.e., full coordination to maximize the sum of welfare). A number of numerical simulations show that the Rosen equilibrium is fairly close to the Pareto optimum while the Nash equilibrium (without a coupled constraint) is much below the welfare-possibility frontier.

---

[27]This formulation corresponds to the concept of two-level hierarchical games proposed by Helm (2003) where, for example, countries decide on their own emissions while agreeing on an international emissions trading scheme.

Chapter 3

DYNAMIC GAMES IN NATURAL
RESOURCES ECONOMICS

This chapter begins with a survey of dynamic games involving renewable resources. This is followed by games involving an exhaustible resource. Finally, we discuss related topics such as the strategic interactions among players who use antibiotics or pesticides, knowing that bacteria and insects gradually develop resistance.

3.1. Renewable Resources

Renewable resources are natural resources that can potentially be maintained at positive steady-state levels if harvesting is carefully managed. Without proper management, however, their exhaustion (or extinction) is a definite possibility. Forests, aquafiers, fish species, and animal species are common examples of renewable resources. The long-term sustainability of many renewable resources is threatened by excessive exploitation, partly because of lack of co-operation among people who have access to them. This problem is known as the tragedy of the commons. In the next section, we turn to some dynamic game models that shed light on this problem.

3.1.1. *The tragedy of the commons*

The tragedy of the commons has long been well recognized (Gordon, 1954; Hardin, 1968). If individuals have common access to a resource stock, they tend to overexploit it. While some societies succesfully develop institutions and norms of behavior that to some extent mitigate the tragedy of the commons (Ostrom, 1990), there are obvious cases of extreme overexploitation. The Food and Agriculture Organization reported that, in 2007, 80% of the stocks are fished at or beyond their maximum sustainable yield (FAO, 2009). Grafton *et al.* (2007) documented serious overexploitation of several fish species. Recent empirical work by

McWhinnie (2009) found that shared stocks are indeed more prone to overexploitation, confirming the theoretical prediction based on a dynamic game model of Clark and Munro (1975), that an increase in the number of players reduces the equilibrium stock level.

This section provides a brief survey of models of common access fishery. The "fishery model" has been interpreted more broadly to mean a model about almost any kind of renewable resource. This interpretation has been taken by many authors, for example, Brander and Taylor (1997) and Copeland and Taylor (2009). Using a similar framework, Weitzman (1974b) compares free access versus private ownership as alternative systems for managing common property, Chichilnisky (1994) considers a North-South trade model with common property resources. These four papers do not deal with situations involving dynamic games.

Inefficiency and overharvesting results: Some open-loop illustrations

Clark and Munro (1975) consider a dynamic game among n infinitely lived fishermen who have access to a common fish stock, denoted by $S(t)$. Below is a generalized version of their model. The world price of fish is an exogenous constant $p > 0$. Each fisherman's effort is denoted by $L_i(t)$. The opportunity cost of effort is $c > 0$ per unit (which represents the wage he can earn in an alternative occupation). At each t, there is an upper bound $\overline{L}$ on his effort. His catch at t is $h_i(t) = qL_i(t)(S(t))^{\mu}$ where $q > 0$ is the catchability coefficient, and $\mu \geq 0$ is the elasticity of harvest with respect to stock. (This function is known as the Schaefer harvest function when $\mu = 1$, see Schaefer, 1957.) The rate of change in the fish stock is given by

$$\dot{S}(t) = G(S) - \sum_{i=1}^{n} qL_i(t)(S(t))^{\mu}$$

where $G(S)$ is the natural growth function. An example of $G(S)$ is

$$G(S) = rS(t)\left(1 - \frac{S(t)}{K}\right)$$

Here $r > 0$ is the proportional rate of growth of the stock, $\dot{S}/S$, if S is near zero, and $K > 0$ is called the carrying capacity: when $S > K$, the stock will decline even in the absence of harvesting.

Clark and Munro describe the open-loop Nash equilibrium. Let us present a slightly more general version of their model. Fisherman i's profit is

$$\pi_i(t) = pqL_i(t)(S(t))^\mu - cL_i(t)$$

Assume the utility of profit is

$$U(\pi_i) = \frac{\pi_i^{1-\alpha}}{1-\alpha}$$

where $\alpha \geq 0$. Clark and Munro (1975) consider the case where $\alpha = 0$.

Fisherman i takes as given the time path of effort of all other agents and chooses his own time path $L_i(t) \in [0, \overline{L}]$ to maximize the integral of discounted utility

$$V_i = \int_0^\infty e^{-\delta t} \frac{1}{1-\alpha} [\pi_i(t)]^{1-\alpha} dt$$

subject to the transition equation, the initial condition $S(0) = S_0$ and the terminal condition $\lim_{t\to\infty} S(t) \geq 0$. Here, δ is the positive discount rate. This model can be generalized further, as in Long and McWhinnie (2010), where it is assumed that fishermen care about relative income, that is,

$$U = \frac{1}{1-\alpha} (\pi_i - \gamma\overline{\pi})^{1-\alpha}$$

where $\overline{\pi}$ is the average profit of one's peer group and $0 \leq \gamma \leq 1$.[1]

The externality in this model is that each agent does not take into account the fact that his effort today will raise other fishermen's costs tomorrow via its effect on tomorrow's stock. It can be shown that, assuming an interior solution in the steady-state, the symmetric open-loop Nash equilibrium results in a steady-state S_∞^{OL} that satisfies the following "externality-distorted modified golden rule":

$$\delta = G'(S_\infty^{OL}) + \frac{\mu}{n} \left(\frac{G(S_\infty^{OL})}{S} \right) \left[\frac{c}{pq(S_\infty^{OL})^\mu - c} - (n-1) \right] \qquad (3.1)$$

If the fishermen co-operate and coordinate their effort to maximize the sum of their V_i, the resulting (optimal) steady-state stock, denoted by S_∞^{so} will be higher than S_∞^{OL}, and everyone will be better off. In fact, S_∞^{so} satisfies

---

[1]They show that to achieve efficiency, two taxes are required: a tax on relative profit, and a tax on output.

the following "modified golden rule"

$$\delta = G'(S_\infty^{so}) + \mu \left(\frac{G(S_\infty^{so})}{S} \right) \left[\frac{c}{pq(S_\infty^{OL})^\mu - c} \right] \qquad (3.2)$$

Notice that if $n = 1$, then the two Eqs. (3.1) and (3.2) would be identical. The first term on the RHS of Eq. (3.2) is the marginal natural growth rate of the stock (called the biological rate of interest) and the second term is the marginal benefit (in terms of cost reduction) of keeping an extra fish in the pool: it is equal to the *group* steady-state harvest per unit of stock, multiplied by the (per unit of effort) cost/profit ratio, and the elasticity of harvest with respect to stock, μ. In contrast, in the second term on the RHS of Eq. (3.1), only the *individual* steady-state harvest is counted in the marginal benefit term, and the steady-state harvest of the other $n - 1$ agents is considered as a reduction in the individual's rate of return in leaving an additional fish in the pool.

It is important to note that the distortion arises because the stock S has an effect on the catch, that is, $\mu > 0$. If the catch depends only on effort and is independent of S (i.e., $\mu = 0$) then there is no externality in the steady state. In fact, Chiarella *et al.* (1984) show that when costs are stock independent in a multi-species model there exist open-loop Nash equilibria that are Pareto efficient (and other OLNE that are not Pareto efficient, see their appendix). Similar results using open-loop Nash equilibrium are reported in Dockner and Kaitala (1989). However, as shown by Martin-Herrán and Rincón-Zapareto (2005), if agents in the model of Chiarella *et al.* (1984) use continuous feedback strategies, the resulting feedback Nash equilibrium is always inefficient.[2]

Feedback Nash equilibrium in a common access fishery

As we have pointed out in earlier chapters, the concept of open-loop Nash equilibrium is based on the assumption that each player is committed to a given time path of action. This may not seem reasonable. If for some reason the stock level at a given time deviates from what everyone anticipated, clearly players will no longer want to stick to the committed path.

---

[2]It is important to recall the Markov-perfect requirement in the definition of a feedback Nash equilibrium: starting at *any* (state, date) pair which may result from *trembling-hand* deviations, each agent's equilibrium strategy (decision rule) must maximize his continuation payoff, given the equilibrium strategies of other players.

Feedback Nash equilibria have been explored in fishery games, both in discrete time and in continuous time. In general, multiplicity of feedback equilibria may arise, as shown in Dockner and Sorger (1996), and Sorger (1998).[3]

One of the earliest studies of feedback equilibrium in dynamic exploitation of a common property resource is the fish-war model of Levhari and Mirman (1980), formulated in discrete time. Given below is a version of their model.[4]

Consider a stock of fish, denoted by s_t, shared by two countries, say country i and country j. The harvest rates are c_t^i and c_t^j and the transition equation is

$$s_{t+1} = (s_t - c_t^i - c_t^j)^\alpha, \quad 0 < \alpha < 1$$

where $c_t^i + c_t^j \le s_t$. The utility derived by country k is $\ln(c_t^k)$.

Let us look at the co-operative scenario. The joint optimization problem is

$$\max \sum_{t=0}^{\infty} \beta^t \left[\ln(c_t^i) + \ln(c_t^j) \right]$$

Clearly, given the strict concavity of the objective function, the two countries would choose the same rate of exploitation $c_t^i = c_t^j = \widetilde{c}_t$. Then, the joint maximixation problem becomes that of choosing the time path $\widetilde{c}_t$ to maximize

$$\max \sum_{t=0}^{\infty} 2\beta^t \ln \widetilde{c}_t$$

subject to

$$s_{t+1} = (s_t - 2\widetilde{c}_t)^\alpha$$

Define the aggregate harvest by

$$h_t = 2\widetilde{c}_t$$

---

[3]For fundamental results on the existence of feedback equilibrium, see Sundaram (1989) and Dutta and Sundaram (1992).
[4]Takayama and Simaan (1984) also analyze a dynamic game involving two countries and a single fish stock.

Then, we seek

$$\max \sum_{t=0}^{\infty} 2\beta^t \ln \left[\frac{h_t}{2}\right] = \max \sum_{t=0}^{\infty} 2\beta^t \left[\ln h_t - \ln 2\right]$$

Let $W(s)$ be the value function. Then, the (discrete-time) Bellman equation for the joint maximization problem is

$$W(s_{t+1}) = \max \left\{2\ln h_t - 2\ln 2 + \beta W[(s_t - h_t)^{\alpha}]\right\}$$

Let us conjecture that the value function takes the form

$$W(s) = K \ln s + M$$

Proceeding in the usual way, we get the optimal aggregate harvest rule

$$h_t = (1 - \beta\alpha)s_t$$

This exploitation rule results in a time path of the stock, s_t, that converges to a unique steady state

$$\widehat{s} = (\beta\alpha)^{\alpha/(1-\alpha)}$$

Now, suppose the two countries do not co-operate. Let us find the Markov-perfect Nash equilibrium of this game. Now, country i thinks that country j uses the harvesting strategy

$$c_t^j = \phi^j(s_t)$$

where $\phi^j(0) = 0$ and $\phi_s^j(s) > 0$. Country i takes $\phi^j(.)$ as given, and finds the path c_t^i to maximize

$$\sum_{t=0}^{\infty} \beta^t \ln c_t^i$$

subject to

$$s_{t+1} = (s_t - \phi^j(s_t) - c_t^i)^{\alpha}$$

The Bellman equation for country i is

$$V^i(s_{t+1}) = \max_{c_t^i} \left\{\ln c_t^i + \beta V^i[(s_t - \phi^j(s_t) - c_t^i)^{\alpha}]\right\}$$

Suppose country i thinks that country j uses the following linear havest strategy

$$\phi^j(s) = \gamma^j s \quad \text{where } 0 < \gamma^j < 1$$

Let us conjecture that the value function of country i takes the form

$$V^i(s) = A_i \ln s + B_i$$

Then

$$A_i \ln s + B_i = \max_{c^i}\{\ln c^i + \beta A_i \alpha \ln(s - \gamma^j s - c^i) + \beta B_i\}$$

Maximizing the RHS with respect to c^i gives the FOC

$$\frac{1}{c^i} = \frac{\beta \alpha A_i}{(1 - \gamma^j)s - c^i} \tag{3.3}$$

So, the Bellman equation becomes

$$A_i \ln s + B_i = \ln s + \ln\left(\frac{1 - \gamma^j}{1 + \beta \alpha A_i}\right)$$

$$+ \beta \alpha A_i \ln s + \beta \alpha A_i \ln\left[\frac{(1 - \gamma^j)\beta \alpha A_i}{1 + \beta \alpha A_i}\right] + \beta B_i$$

Hence

$$[1 - A_i(1 - \beta \alpha)]\ln s = B_i(1 - \beta) - \ln\left(\frac{1 - \gamma^j}{1 + \beta \alpha A_i}\right)$$

$$- \beta \alpha A_i \ln\left[\frac{(1 - \gamma^j)\beta \alpha A_i}{1 + \beta \alpha A_i}\right]$$

This equation must hold for all $s > 0$. It follows that the coefficient of the term $\ln s$ must be zero, that is,

$$A_i = \frac{1}{1 - \beta \alpha} \tag{3.4}$$

Substituting (3.4) into the FOC (3.3) yields player i's feedback strategy

$$c^i = (1 - \beta \alpha)(1 - \gamma^j)s \equiv \gamma^i s$$

This gives γ^i to be a function of γ^j

$$\gamma^i = (1 - \beta \alpha)(1 - \gamma^j) \quad \text{for } 0 \le \gamma^j \le 1$$

This equation may be interpreted as player i's "reaction function". It is downward sloping, with slope $-(1 - \beta \alpha)$. Thus, the higher is country j's harvesting effort, the lower will be the input level of country i. (Efforts are strategic substitutes.)

By a similar argument applied to country j, we obtain

$$c^j = (1 - \beta\alpha)(1 - \gamma^i)s \equiv \gamma^j s$$

and

$$\gamma^j = (1 - \beta\alpha)(1 - \gamma^i) \quad \text{for } 0 \leq \gamma^i \leq 1$$

The two "reaction functions" have a unique intersection, denoted by γ, where

$$\gamma = \frac{1 - \beta\alpha}{2 - \beta\alpha} \tag{3.5}$$

This value of γ gives us the symmetric feedback equilibrium. Let us consider the transition equation under this feedback equilibrium

$$s_{t+1} = [s_t - 2\gamma s_t]^\alpha = \left(\frac{\beta\alpha}{2 - \beta\alpha}\right)^\alpha s_t^\alpha$$

The steady-state biomass is then

$$\tilde{s} = \left(\frac{\beta\alpha}{2 - \beta\alpha}\right)^{\alpha/(1-\alpha)}$$

which is smaller than the steady-state stock under co-operation. This confirms that the tragedy of the commons can occur even if the harvesting function depends only on the effort and is independent of the stock. This result contrasts sharply with the open-loop result found by Chiarella *et al.* (1984). See also the subsection below on linear feedback strategies in continuous time.

The discrete time model of Levhari and Mirman (1980) has been generalized by Antoniadou *et al.* (2008) to the case where the utility function is

$$u(c) = \frac{c^{\frac{\eta-1}{\eta}} - 1}{\frac{\eta-1}{\eta}}$$

where $\eta > 0$ is the intertemporal rate of substitution, and the transition equation is

$$s_{t+1} = \theta \left[\alpha(y_t)^{\frac{\eta-1}{\eta}} + (1-\alpha)k^{\frac{\eta-1}{\eta}} \right]^{\frac{\eta}{\eta-1}}$$

where $y_t = s_t - \sum c_{it}$ and k is a constant.[5] The limiting case where $\eta \to 1$ is the original Levhari-Mirman paper (1980). Note the same parameter η appears in both functions. This special configuration allows the existence of a Markov-perfect equilibrium with linear feedback strategies. When θ is a random variable, Antoniadou *et al.* (2008) show that increases in risks do have an effect of the equilibrium feedback strategy (except in the case where $\eta = 1$). If the support of the shock θ is unbounded, conditions must be placed on the distribution of the shock to ensure existence of value functions[6]. Assuming that $\ln \theta$ is normally distributed with mean $\mu - \sigma^2/2$ and variance σ^2 (so that $E\theta = e^\mu$ and $\mathrm{var}(\theta) = e^{2\mu}(e^{\sigma^2} - 1)$), Antoniadou *et al.* (2008) show that an increase in σ will amplify (respectively, mitigate) the tragedy of the commons if $\eta < 1$ (respectively, $\eta > 1$). Another direction of generalization is to allow for several fish stocks having biological externalities, see Datta and Mirman (1999).

For additional contributions of dynamic games involving renewable resources, see Fischer and Mirman (1986), Benhabib and Radner (1992), Crabbé and Long (1993), Mason and Polasky (1994), Dockner and Sorger (1996), Benchekroun and Long (2002a), Benchekroun (2003, 2008), Fujiwara (2009), among others. The possibility of enforcing co-operation through the use of punishment activated by trigger strategies is discussed and analyzed by Dockner *et al.* (1996). Problems of existence of an MPNE (with or without random disturbance) are discussed by Dutta and Sundaram (1993a,b) and Duffie *et al.* (1994).

3.1.2. *Can Nash equilibria in renewable-resource exploitation be Pareto efficient?*

In repeated games, it is well known that if players use trigger strategies to punish deviation from an allocation, non-cooperative players can achieve Pareto-efficient outcomes, provided that they do not discount the future payoffs too heavily. In dynamic games, where state variables are evolving over time, trigger strategies can also serve to achieve a co-operative outcome (see e.g., Haurie and Pohjola, 1987; Benhabib and Radner, 1992).Trigger strategies require memory of the history of the play, and therefore are

---

[5]When the number of players is 1, this model reduces to the case studied by Benhabib and Rustichini (1994).
[6]See Stachurski (2002) for sufficient conditions on the distribution of θ in the case of a single player.

not Markovian. Another route to achieve efficiency is via the concept of "incentive equilibrium", see for example, Ehtamo and Hämäläinen (1989, 1993) and Jørgensen and Zaccour (2001b).[7]

Martin-Herrán and Rincón-Zapareto (2005) derive necessary conditions for an MPNE to be Pareto efficient, and apply these conditions to a fishery game proposed by Clemhout and Wan (1985). In this game, despite common access, under certain parameter values, an MPNE can be efficient if each player derives pleasure from other players' consumption. This is not surprising: the positive externalities of altruism and the negative externalities of common access just happen to cancel each other out. It is instructive to look at the mechanism that drives this result. Suppose there are two players, and a single fish stock denoted by $s(t)$. Let $c^i(t)$ be the harvest (and consumption) by player i. Assume that time is continuous, and the transition equation takes the form

$$\dot{s}(t) = s(t)[1 - b\ln(s(t))] - c^1(t) - c^2(t), \quad b > 0$$

The instantaneous reward (or utility) that player i gets at time t is

$$R^i(t) = w^i \ln(s) + \mu^i \ln c^i(t) + \gamma^i_j \ln c^j(t), \quad w^i > 0, \quad \mu^i > 0$$

where γ^i_j represents the effect of j's consumption on i's instantaneous utility. The parameter w^i indicates the strength of i's pleasure derived directly from the stock (e.g., the amenity benefits), and the parameter μ^i indicates the intensity of pleasure derived from one's own consumption. If the parameter γ^i_j is positive, we say i is altruistic, if it is negative, we say that i is envious. Player i's payoff is defined to be the integral of the stream of instantaneous reward (from time zero to time infinity), discounted at the rate ρ^i:

$$J^i \equiv \int_0^\infty e^{-\rho^i t} R^i(t) dt$$

Martin-Herrán and Rincón-Zapareto (2005) consider two cases.

---

[7]The concept of "incentive equilibrium" is based on the assumption that player i can threaten player j such that if j deviates from a co-operative path, i will punish j by behaving like a kind of follower in a static leader–follower game. It is not clear whether such punishment is fully rational.

Case 1: Assume that $\rho^1 = \rho^2 = \rho$ and preferences are symmetric. If the following relationships happen to hold

$$\gamma_2^1 = \frac{w^1}{w^2}\mu^2 \quad \text{and} \quad \gamma_1^2 = \frac{w^2}{w^1}\mu^1$$

then one can show that the following pair of harvesting strategies constitutes a Pareto-efficient MPNE

$$c^1 = \frac{(b+\rho)\mu^1 w^2}{\mu^1 w^2 + \mu^2 w^1 + w^1 w^2}x$$

$$c^2 = \frac{(b+\rho)\mu^2 w^1}{\mu^1 w^2 + \mu^2 w^1 + w^1 w^2}$$

Case 2: Assume that $\rho^1 = \rho^2 = \rho$ and preferences are asymmetric. In particular, assume $\gamma_2^1 > 0$ but γ_1^2 may be negative (i.e., agent 1 is altruistic but agent 2 may be envious), but not too negative, so that

$$w^2 + \gamma_1^2 > 0$$

If the following relationship happens to hold

$$\gamma_2^1 = \frac{(w^1 + \mu^1)\mu^2}{w^2 + \gamma_1^2}$$

then one can show that following pair of harvesting strategies constitutes a Pareto-efficient MPNE

$$c^1 = \frac{(b+\rho)\mu^1(w^2 + \gamma_1^2)}{(\mu^1 + w^1)(\mu^2 + w^2 + \gamma_1^2)}x$$

$$c^2 = \frac{(b+\rho)\gamma_2^1(w^2 + \gamma_1^2)}{(\mu^1 + w^1)(\mu^2 + w^2 + \gamma_1^2)}x$$

The model can be generalized to the case with two fish stocks s_1 and s_2 that bear certain predator–prey relationships. Let c_k^j denote player j's harvest from stock k. Assume that the transition equations are

$$\dot{s}_1 = s_1[1 - b_1 \ln s_1 - g_2^1 \ln s_2] - c_1^1 - c_1^2$$
$$\dot{s}_2 = s_2[1 - b_2 \ln s_2 - g_1^2 \ln s_2] - c_2^1 - c_2^2$$

and the instantaneous utility of player i is

$$R^i = w_1^i \ln s_1 + w_2^i \ln s_2 + \mu_1^i \ln c_1^i + \mu_2^i \ln c_2^i + \gamma_{21}^i \ln c_1^j + \gamma_{22}^i \ln c_2^j$$

Again, Martin-Herrán and Rincón-Zapareto (2005) are able to find conditions on parameter values such that there are Pareto-efficient MPNE.

It is interesting to note that by a suitable transformation of variables the problem is linear in the state variables and falls into the class of games called "linear-state games" for which solutions are easily obtained (see Dockner *et al.*, 2005). Consider the following transformation for stock h, k ($h, k = 1, 2$):

$$y_h = \ln s_h, \quad h = 1, 2$$

Then

$$\dot{y}_h = \frac{1}{s_h}\dot{s}_h = 1 - b_h y_h - g_k^h y_k - \frac{c_h^1}{s_h} - \frac{c_h^2}{s_h}$$

Next, define

$$d_h^i = \frac{c_h^i}{s_h}$$

Then player i's instantaneous utility R^i is linear in y_h and y_k. The implications of this observation in terms of value functions and strategies are taken up further in the next subsection.

3.1.3. *Some technical notes on feedback strategies in fishery problems*

Feedback strategies in fishery problems are, in general, difficult to compute analytically. Analytical solutions turn out to be easy to find if each player's problem can be transformed into an optimization problem that is linear in the (transformed) state variable in such a way that the first-order condition with respect to the control variable is independent of the state variable. Let us illustrate. Consider first the Levhari-Mirman model (1980) discussed above. Suppose that each player chooses a linear decision rule $c_t^i = \gamma_t^i s_t$ where $\gamma_t^i \in (0,1)$. Then, the transition equation is

$$\ln s_{t+1} = \alpha \ln s_t + \alpha \ln(1 - \gamma_t^i - \gamma_t^j)$$

and using the transformation of variable

$$z_t = \ln s_t$$

we see that the transition equation is linear in z_t

$$z_{t+1} = \alpha z_t + \alpha \ln(1 - \gamma_t^i - \gamma_t^j)$$

Similarly, the objective function of player i becomes

$$\max \sum_{t=0}^{\infty} \beta^t [z_t + \ln \gamma_t^i]$$

The Bellman equation of player i then becomes linear in the new state variable z:

$$W_i(z_t) = \max_{\gamma_t^i} \{z_t + \ln \gamma_t^i + \beta W_i(\alpha z_t + \alpha \ln(1 - \gamma_t^i - \gamma_t^j))\} \qquad (3.6)$$

Suppose player i thinks that player j uses the constant strategy $\gamma_t^j = \gamma^j$. Then the maximization of the RHS of Eq. (3.6) gives the first-order condition

$$\frac{1}{\gamma_t^i} = \beta W_i'(z_{t+1}) \frac{\alpha}{1 - \gamma_t^i - \gamma^j} \qquad (3.7)$$

Since the Bellman equation is linear in z, we conjecture that $W(z) = H + Kz$, such that $W'(z) = K$, a constant. Then, Eq. (3.7) gives γ_t^i as a constant

$$\frac{1}{\gamma^i} = \frac{\alpha \beta K}{1 - \gamma^i - \gamma^j}$$

that is, the first-order condition is independent of the state variable:

$$\gamma^i = \frac{1}{1 + \alpha \beta K}(1 - \gamma^j) \qquad (3.8)$$

Assuming symmetry, we then have

$$\gamma = \frac{1}{2 + \alpha\beta K}$$

Substituting this into the Bellman equation, we get

$$H + K z_t = z_t - \ln(2 + \alpha\beta K) + \beta H$$
$$+ \beta K \left[\alpha z_t + \alpha \ln \left(1 - \frac{2}{2 + \alpha\beta K} \right) \right]$$

This holds only if

$$K = 1 + \alpha\beta K$$

that is,

$$K = \frac{1}{1 - \alpha\beta}$$

Hence

$$\gamma = \frac{1}{2 + \alpha\beta \left(\frac{1}{1-\alpha\beta} \right)} = \frac{1 - \alpha\beta}{2 - \alpha\beta}$$

which is identical to Eq. (3.5).

A similar feature applies to models in continuous time: a pair of linear harvest rules, $c^i = \gamma^i s$, $i = 1, 2$, constitutes MPNE strategies if by a suitable transformation of variables, the value function is linear in the transformed state variable. Let us illustrate. Suppose the stock of fish is $x(t)$ and its net growth rate is

$$\dot{x}(t) = A x(t)^\theta - \delta x(t) - c_1(t) - c_2(t), \quad A > 0$$

where c_i is the catch rate of country i. Assume $0 < \theta < 1$, $\delta > 0$ and $A > 0$. This implies that the natural growth function $A x(t)^\theta - \delta x(t)$ is *strictly concave and its graph has the inverted U-shape*. Since the derivative of this function, when evaluated at $x = 0$, is infinite, we can be sure that, starting with any $x(0) > 0$, the steady-state stock will be positive.

Assume the utility function is

$$U(c_i) = \frac{c_i^{1-\beta}}{1 - \beta}$$

Now, make the very special assumption that $\beta = \theta$. Consider the transformation of the state variable

$$X(t) \equiv x(t)^{1-\theta}$$

and of the control variables by defining the variable $\omega_i(t)$ as the catch rate per unit of stock

$$c_i(t) = \omega_i(t)x(t)$$

Then

$$\dot{x}(t) = Ax(t)^{\theta} - (\delta + \omega_1(t) + \omega_2(t))x(t) \tag{3.9}$$

and (omitting the time argument)

$$\dot{X} = (1-\theta)x^{-\theta}\dot{x} = (1-\theta)x^{-\theta}[Ax^{\theta} - (\delta + \omega_1 + \omega_2)x] \tag{3.10}$$
$$= (1-\theta)A - (1-\theta)(\delta + \omega_1 + \omega_2)X$$

Player i chooses the time path of ω_i to maximize

$$\int_0^{\infty} e^{-\rho t} \left[\frac{(\omega_i(t)x(t))^{1-\beta}}{1-\beta} \right] dt$$

subject to the transition equation (3.10). Now, if we assume $\beta = \theta < 1$ then we obtain an optimization problem that is linear in the state variable

$$\max_{\omega_i} \int_0^{\infty} e^{-\rho t} \left[\frac{(\omega_i(t))^{1-\theta}}{1-\theta} X(t) \right] dt$$

subject to (3.10).

Suppose player i thinks that player j uses a constant harvesting strategy, $\omega_j(t) = \omega_j$. The HJB equation of player i is then

$$\rho V_i(X) = \max \left\{ \frac{X\omega_i^{1-\theta}}{1-\theta} + V_i'(X)\left[(1-\theta)A \right. \right.$$

$$\left. \left. -(1-\theta)(\delta + \omega_i + \omega_j)X\right] \right\}$$

The maximization of the RHS yields

$$X\omega_i^{-\theta} - X(1-\theta)V_i'(X) = 0$$

Notice that this implies that $\omega_i(t)$ is independent of ω_j. Let us conjecture that the value function is linear in X, that is, $V_i(X) = E_i + K_i X$, where $K_i > 0$. Then, the FOC is independent of the state variable:

$$\omega_i^{-\theta} = (1-\theta)K_i$$

Substituting into the HJB equation, and assuming symmetry

$$\rho E + \rho K X = \frac{X}{1-\theta}[(1-\theta)K]^{(\theta-1)/\theta}$$
$$+(1-\theta)AK + (1-\theta)KX[\delta + 2[(1-\theta)K]^{-1/\theta}]$$

we can then determine K and E.

Consider next the case of a natural growth function that has a finite derivative at $x = 0$. Suppose

$$\dot{x}(t) = rx(t) - Bx(t)^\eta - c_1(t) - c_2(t)$$

where $r > 0$, $B > 0$, and $\eta > 1$. (The case $\eta = 2$ is the standard logistic growth function in the fishery literature.) In this case, define the variable X by

$$X(t) \equiv x(t)^{1-\eta}$$

Notice that since $1 - \eta < 0$, a higher value of X means a lower value of the true fish stock x. As $X \to \infty$, $x \to 0$. So, we expect the shadow price of X to be negative. Again, writing $c_i(t) = \omega_i(t)x(t)$, we get

$$\dot{X} = (\eta-1)B + (\eta-1)(\omega_1 + \omega_2 - r)X$$

Assume the utility function

$$U(c_i) = \frac{c_i^{1-\beta}}{1-\beta}$$

Again, take the very special case where $\beta = \eta$. Since $\eta > 1$, this means the utility function is bounded above by zero. Conjecture the value function $V_i(X) = E_i - F_i X$ where $F_i > 0$. This means $V_i'(X) = -F_i < 0$ as we would expect since high X means low x. The HJB equation for player i is

$$\rho V_i(X) = \max_{\omega_i} \left\{ \frac{X\omega_i^{1-\eta}}{1-\eta} + V_i'(X)\left[(\eta-1)B \right.\right.$$

$$\left.\left. +(\eta-1)(\omega_i + \omega_j - r)X\right] \right\}$$

The maximization of the RHS with respect to ω_i yields

$$X\omega_i^{-\eta} + X(\eta-1)V_i'(X) = 0$$

that is, the optimal ω_i is independent of the stock:

$$\omega_i^{-\eta} = (\eta-1)F_i > 0$$

The reader is invited to determine the value of F_i in the case of symmetric agents. (Note that in this case, even with just one player, the steady-state stock of fish x can be zero (i.e., $X \to \infty$) if the rate of discount ρ exceeds the intrinsic growth rate r of the (true) stock x.)

To summarize, to have Markov-perfect equilibrium harvesting strategies of the form $c_i = \gamma x$ where $\gamma > 0$, there must be a special relationship between the natural growth function and the utility function. This feature was stated by Clemhout and Wan (1985), Ploeg (1986, 1987), and further generalized by Gaudet and Lohoues (2008).

Koulovatianos (2007) has explored a more general formulation where there are several fish species involved in a predator–prey relationship. Assume there are two species. Their stocks are denoted by $x(t)$ and $y(t)$. The aggregate harvest rates from these stocks are denoted by $c_x(t)$ and $c_y(t)$.

The rates of growth of the stocks are

$$\dot{x}(t) = A_x x(t)^\theta - \left\{ \delta_x + D_x \left[\frac{y(t)}{x(t)} \right]^{1-\theta} \right\} x(t) - c_x(t)$$

$$\dot{y} = A_y y(t)^\theta - \left\{ \delta_y + D_y \left[\frac{x(t)}{y(t)} \right]^{1-\theta} \right\} y(t) - c_y(t)$$

Assume $0 < \theta < 1, \delta_x > 0, \delta_y > 0$.

If $D_x > 0$ and $D_y > 0$, the two species are said to be competing species. If $D_x < 0$ and $D_y > 0$, we say the first species are predators and the second are preys. Consider for the moment the special case where

$$c_x(t) = b_x x(t)$$

and

$$c_y(t) = b_y y(t)$$

where b_x and b_y are positive constants.

Consider the following transformation of variables

$$X(t) \equiv x(t)^{1-\theta} \quad \text{and} \quad Y(t) \equiv y(t)^{1-\theta}$$

Then, we obtain the matrix equation

$$\dot{s} = A + Bs \qquad (3.11)$$

where

$$s \equiv \begin{bmatrix} X \\ Y \end{bmatrix}, \quad A \equiv (1-\theta) \begin{bmatrix} A_x \\ A_y \end{bmatrix}$$

$$B \equiv (1-\theta) \begin{bmatrix} -\delta_x - b_x & -D_x \\ -D_y & -\delta_x - b_y \end{bmatrix} \equiv \begin{bmatrix} b_{11} & b_{12} \\ b_{21} & b_{22} \end{bmatrix}$$

Note that $tr(B) < 0$. Assume that $Det(B) > 0$. (For example, D_x and D_y are sufficiently small in absolute value.) Then both eigenvalues, denoted by μ_1 and μ_2, are negative. We will assume that $\mu_1 \neq \mu_2$. Then, the system has a unique steady state:

$$s^{\text{ss}} = -B^{-1}A$$

In particular

$$X^{\text{ss}} = \frac{1-\theta}{(\delta_x + b_x)(\delta_y + b_y) - D_x D_y} [A_x(\delta_y + b_y) - D_x A_y]$$

Since x and y cannot be negative, we must choose parameter values such that X^{ss} and Y^{ss} are positive, for example, by assuming the coefficients D_x and D_y are small enough. Uncertainty can be added without much complication, as in Koulovatianos (2007).

3.1.4. *Differential game models of an oligopolistic fishery*

The above fishery models are based on the assumption that individual fishermen have no impact on the market price. This assumption is relaxed in a number of papers, including Dockner *et al.* (1989), Jorgensen and Yeung (1996), Benchekroun (2003, 2008), Lohoues (2006), Fujiwara (2009a,b).[8] Benchekroun (2003) assumes an inverted V-shaped natural growth function and a linear demand function, and shows that an exogenous unilateral restriction in one firm's harvest can lead to a decrease in the steady-state stock. Benchekroun (2008) shows that an increase in the number of firms results in a lower steady-state industry output.

Jorgensen and Yeung (1996) assume that the evolution of the fish stock follows the following stochastic differential equation

$$dx = \left[ax^{1/2} - bx - \sum_{i=1}^{N} h_i \right] dt + \sigma x dW \qquad (3.12)$$

where W is a Wiener process, that is, dW is normally distributed with mean zero and variance σ^2. Here $h_i(t)$ is agent i's harvest at time t. The parameter b is the death rate.

The total amount of fish caught at time t is

$$Q(t) = \sum_{i=1}^{N} h_i(t)$$

It is assumed that the inverse demand function is

$$P = Q^{-1/2}$$

Player i's total cost of catching h_i fish is

$$C(h_i, x) = \frac{c}{\sqrt{x}} h_i$$

---

[8]Sandal and Steinshamm (2004) also consider a Cournot oligopoly in a fishery facing a downward sloping linear demand schedule. However, they assume that at most one firm takes into account the stock dynamics. Myopic behavior is also assumed by Hämäläinen *et al.* (1986, 1990).

Player i seeks to maximize

$$E_0 \left\{ \int_0^\infty e^{-\rho t} \left[Ph_i - \frac{c}{\sqrt{x}} h_i \right] dt \right\}$$

The authors show that the value function of i is

$$V_i(x) = A\sqrt{x} + \frac{(2N-1)c}{4N^3}$$

where A is the unique positive root of the cubic equation

$$\frac{1}{4} z A^3 + z c A^2 + \left[z c^2 + \frac{4N^2 - 8N + 3}{8N^2} \right] A = \frac{(2n-1)c}{4n^3}$$

and where $z \equiv \rho + \sigma^2/8 + b/c$. The equilibrium feedback strategy is

$$h(x) = \left(\frac{(2N-1)c}{4N^3} \right) \left(c + \frac{A}{2} \right)^{-2} x$$

They show that at any given fish stock, an increase in the death rate b or an increase in the variance σ^2 will increase the harvest rate, while an increase in the cost parameter c will reduce the harvest rate.

It is worth noting that the above model can be generalized as follows. Instead of the transition equation (3.12), we specify a more general function

$$dx = \left[ax^\theta - bx - \sum_{i=1}^N h_i \right] dt + \sigma x dW \quad \text{where } \theta \in (0,1) \qquad (3.13)$$

Further, instead of the inverse demand function $P = Q^{-1/2}$, we specify $P = Q^{-(1-\theta)}$ such that the industry's total revenue is $PQ = Q^\theta$. The cost function is

$$C(h_i, x) = \frac{ch_i}{x^{1-\theta}}$$

Without loss of generality, define the harvesting intensity of firm i as

$$\omega_i(t) = \frac{h_i(t)}{x(t)}$$

Then, the instantaneous profit of firm i is

$$\pi_i = Ph_i - \frac{ch_i}{x^{1-\theta}} = \frac{\omega_i x}{\left[\omega_i + \sum_{j \neq i} \omega_j\right]^{1-\theta} x^{1-\theta}} - \frac{c\omega_i x}{x^{1-\theta}}$$

$$= x^\theta \left[\frac{\omega_i}{\left(\omega_i + \sum_{j \neq i} \omega_j\right)^{1-\theta}} - c\omega_i\right] \qquad (3.14)$$

Now, let us transform the state variable by defining

$$Y = x^\theta \equiv F(x)$$

From Itô's lemma[9], if $dx = g(x,t)dt + v(x,t)dW$, then

$$dY = \left[F_x g(x,t) + \frac{1}{2}F_{xx}v^2(x,t)\right] + F_x v(x,t)dW \qquad (3.15)$$

Hence, combining (3.15) and (3.13)

$$dY = \left[\theta x^{\theta-1}\left(ax^\theta - bx - \sum_{i=1}^{N}\omega_i x\right) + \frac{1}{2}\theta(\theta-1)x^{\theta-2}\sigma^2 x^2\right]dt$$

$$+\theta x^{\theta-1}\sigma x dW$$

$$= \left[\theta a - \theta Y\left(b + \frac{1}{2}(1-\theta)\sigma^2 + \sum_{j=1}^{N}\omega_j\right)\right]dt$$

$$+\theta\sigma Y dW \qquad (3.16)$$

The RHS of (3.16) is linear in Y. The instantaneous profit function (3.14) is also linear in Y. Thus we have transformed the generalized model of Jorgensen and Yeung into a differential game that is linear in the state variable Y. The solution is now straightforward. Assume that player i thinks

---

[9]For a brief introduction to Itô's lemma, see Kamien and Schwartz (1991, Section 22).

that all other players behave identically and use a constant harvesting intensity ω_j. Then, Eq. (3.16) becomes

$$dY = k(Y, \omega_i, \omega_j)dt + \theta\sigma Y\, dW$$

where

$$k(Y, \omega_i, \omega_j) = \theta a - \theta Y \left(b + \frac{1}{2}(1 - \theta)\sigma^2 + \omega_i + (N - 1)\omega_j \right)$$

Then, the HJB function for player i is

$$\rho V_i(Y) = \max_{\omega_i} \left\{ \frac{\omega_i Y}{(\omega_i + (N - 1)\omega_j)^{1-\theta}} - c\omega_i Y + V_i'(Y)k(Y, \omega_i, \omega_j) \right.$$
$$\left. + \frac{1}{2}V_i''(Y)(\theta\sigma Y)^2 \right\}$$

Now, we conjecture that the value function is linear in Y:

$$V_i(Y) = AY + B$$

such that $V_i' = A$ and $V_i'' = 0$. Substituting into the HJB equation, and maximizing the RHS with respect to ω_i gives the FOC:

$$\frac{(\omega_i + (N - 1)\omega_j)^{1-\theta} - \omega_i(1 - \theta)(\omega_i + (N - 1)\omega_j)^{-\theta}}{(\omega_i + (N - 1)\omega_j)^{2(1-\theta)}} = \theta A + c$$

In a symmetric equilibrium, $\omega_i = \omega_j = \omega$. So

$$\omega^{1-\theta}N^{1-\theta} = (\theta A + c)\left(\frac{N + 1 - \theta}{N} \right)$$

Solving for ω in terms of N, A, c, and θ, and substituting it into the HJB equation, we can determine A and B. In the special case where $\theta = 1/2$, we get exactly the same A and B as obtained by Jorgensen and Yeung (1996).

3.1.5. *Entry deterrence*

In the models surveyed above, the number of players is exogenously fixed. There are situations where the number of players are endogenously determined, for example when incumbent firms must choose whether to accommodate or to deter entrants. Crabbé and Long (1993) consider a nation whose fishery industry faces foreign poachers. In the case where poachers take the average catch per vessel as given, the country, acting as the Stackelberg leader, can deter the entry of poachers by overfishing, as the reduced stock level raises their harvesting costs. In the case where poachers take a more strategic view (i.e., each knows its impact on the marginal product of all vessels), the Stackelberg leader finds it optimal to accommodate entry, and in the steady state there are active poachers. Social welfare of the country decreases in both cases, as compared with sole ownership. Mason and Polasky (1994) consider a two-period model with a single firm facing potential entry of a rival firm. The incumbent deters entry by increasing its harvest, thus driving down the resource stock to raise the rival's harvesting cost.[10] They show that social welfare can fall as a result of entry deterrence. There is a parallel between this result and the result on "welfare-reducing enclosure" by Long (1994), who shows that an enclosure decision by private owners of properties, which involves a fixed cost per acre, can reduce welfare.

3.1.6. *Further results*

More on common access fishery

For more results on common property games of the fishery type, see Dockner and Sorger (1996), Sorger (1998, 2005), and Benchekroun (2008) on multiplicity of equilibria; Koulovatianos and Mirman (2007) and Antoniadou *et al.* (2008) on linear Markov-perfect equilibria under uncertainty; Amir (1996) shows that for some games an equilibrium does not exist in the deterministic case, while existence is assured for some stochastic versions of the same model.

Intergenerational equity games

The problem of intergenerational equity was raised by Solow (1974) and Hartwick (1977), and further discussed by many authors, including Asheim (1999) and Alvarez-Cuadrado and Long (2009). Long (2006) considers a dynamic game between two governments that are concerned about intergenerational equity. The model is set in discrete time, and adults do

---

[10]The result that renewable-resource firms expand output to deter entry is in sharp contrast with the non-renewable resource case, where an incumbent would increase its initial price (reducing initial output) in response to the threat of a future substitute, leading possibly to lower welfare; see Gilbert and Goldman (1978).

not give bequests to their children. A common property resource stock x_t is harvested in the amounts h_{1t} and h_{2t} by the adults of the two countries. The post-harvest stock, $x_{t+1} = A(x_t - h_{1t} - h_{2t})$, yields the amenity benefit $s_t = \kappa x_{t+1}$. Each country has a government that plans over an infinite horizon and wishes to achieve a stream of constant and positive utility for its citizens. This objective function is called "the sequential maximin objective." Each government imposes a harvest quota rule for its own citizens. Long (2006) shows that there is a continuum of Markov-perfect quota rules, where the quota, expressed as a fraction of the stock, turns out to be a constant. Then, there exists a symmetric Markov-perfect equilibrium for the game between the two governments. The outcome turns out to be identical to the optimum that would be chosen by a fictitious international dictator that has the sequential maximin objective function.

3.2. Non-Renewable Resources

The influence of big oil producers on the price of oil is well documented, and as such the market is oligopolistic, as described by Smith (2009). The list of the top 20 largest oil companies consists mostly of state-owned enterprises of the major oil exporting countries. Not all of these companies are affiliated with the organization of the petroleum exporting countries (OPEC). It seems appropriate to formulate a game of dynamic oligopoly in exhaustible resources.

3.2.1. *Dynamic games with a non-renewable resource: open-loop equilibrium*

One of the earliest models of dynamic games in a market for an exhaustible resource is Salant (1976). He considers (in his Appendix B) an asymmetric oligopoly consisting of $n + 1$ firms, where the last n firms are identical in terms of their stock $\overline{S}$ of an exhaustible resource, while the first firm has a significantly larger stock, in particular, $\overline{S}^{(1)} > n\overline{S}$. Using the open-loop Nash equilibrium concept (i.e., each firm takes the extraction path of all other firms as given), Salant shows that if the demand becomes zero at some finite price, the firm with the largest stock will become a monopolist after all the other firms have exhausted their stocks. Letting n tend to infinity while keeping $n\overline{S}$ constant, Salant (1976) obtains the so-called "Cartel versus Fringe" scenario, where the cartel takes the time path of output of the fringe as given. In this limiting case, extraction will proceed in two phases.

In the first phase, the price (net of marginal extraction cost) rises at the rate of interest, and the market is supplied by the cartel and the fringe. In the second phase, only the large firm remains and it behaves like a monopolist. Ulph and Folie (1980) derive the equilibrium under constant marginal extraction costs, possibly different between the cartel and the fringe.

Salant's dynamic cartel versus fringe model is somewhat different from the static cartel versus fringe model in most textbooks, where the cartel does not take the output of the fringe as given, but rather announces a price to induce an output response from the fringe, that is, the cartel is a Stackelberg leader. Gilbert (1978) therefore considered the case where the cartel has the first mover advantage and the fringe members are followers. A problem with the leader–follower formulation is that in general the open-loop Stackelberg equilibrium is not time consistent, though for the cartel-fringe model there are parameter values such that the open-loop Stackelberg solution is time consistent (Newbery, 1981; Ulph, 1982; Groot *et al.*, 1992). Feedback Stackelberg equilibria have the desirable property of time consistency, but in general it is difficult to characterize them analytically (Groot *et al.*, 2003).

There is an interesting case where oil-well owners have the right to extract the oil located under their own properties, but the oil seeps from one holding to another at a speed dependent on the relative sizes of the stocks currently under each property. Analysis of this problem has been conducted by Khalatbari (1977), Dasgupta and Heal (1979, Ch. 12), Kemp and Long (1980), and Sinn (1983), under open-loop formulation. In the limiting case where the speed of seepage tends to infinity, the model is reduced to the pure common pool problem, analyzed by Bolle (1980). McMillan and Sinn (1984) review the difference among the various open-loop assumptions of the above models, and offer an alternative formulation based on the idea of conjectural variation and ·closed-loop (but not Markov-perfect) decision rules. They find that there are infinitely many equilibria, most resulting in overexploitation, but one resulting in socially optimal extraction. However, none of these is a rational expectation equilibrium.

Salant's model of open-loop asymmetric oligopoly, without common access or seepage, has been generalized by Lewis and Schmalensee (1980), Loury (1986), Gaudet and Long (1994), Benchekroun and Withagen (2008), Benchekroun, Halsema, and Withagen (2009, 2010).

Loury (1986) considers an extractive industry consisting of n non-identical firms, that have the same marginal cost but different deposit sizes.

Under a general demand function, he shows that in an open-loop Nash equilibrium, the average and marginal return on resource stocks is inversely related to the initial stock sizes, that the firms with smaller stocks exhaust their deposits earlier than firms with larger stocks, and that industry production maximizes a weighted average of intertemporal consumer's surplus and profits. Lewis and Schmalensee (1980) allow firms to have different extraction costs. They show that in an open-loop Nash equilibrium the least cost deposit is not necessarily exploited first, resulting in social inefficiency. Polasky (1992) develops the model further and carries out an empirical test. The results on open-loop oligopoly in a non-renewable resource have been generalized further by Benchekroun *et al.* (2009, 2010), where there are two groups of firms, and firms can differ across groups both in their resource stocks and their constant marginal costs. Assuming linear demand and open-loop strategies, they find that there almost always exists a phase where both types of firms have positive extraction rates, that when the high-cost mines are exploited by the group of firms whose number goes to infinity, the equilibrium approaches the cartel-versus-fringe model. They show that the cheaper deposits may not be exhausted first, and that an increase in the reserves of the (higher cost) fringe may lower social welfare. This result is consistent with that obtained by Lahiri and Ono (1988), and generalized by Long and Soubeyran (2001), in a static asymmetric Cournot model, where a small reduction in the cost of a higher cost firm may reduce welfare, because this may increase the market share of the higher cost firm.

Gaudet and Long (1994) consider a marginal transfer of the initial resource stocks from one firm to another in an open-loop duopoly. They find that if the stock sizes are sufficiently different so that the firms have different exhaustion times, then a marginal transfer that makes the distribution more uneven will increase industry profits. On the other hand, if the stocks are not sufficiently different in size, firms will exhaust at the same time, and in that case a marginal transfer of initial stocks has no effect on the industry's output path. This is a generalization of the static model of Bergstrom and Varian (1985).

The open-loop equilibria considered above were obtained in a framework of complete certainty. What happens if there is uncertainty? Kemp and Long (1978) consider the case where firms are uncertain about the size of their stock[11]. Then even with ex-ante identical firms, each firm

---

[11]This type of uncertainty was modeled, in the central planner case, by Kemp (1976, 1977) and has been generalized by Kemp and Long (1980a, 1985).

would take into account the possibility that one day it may become the only firm that still has a positive resource stock, at which time it would change its status to monopoly and follow a monopolistic extraction strategy. In the absence of a complete set of state-contingent markets, the meaning of perfect competition is not clear. Kemp and Long (1978) show that some form of price-taking behavior, accompanied by the recognition that one may suddenly become a monopolist, may result in an equilibrium that is worse than monopoly.

As we have stated before, while the concept of open-loop equilibrium has the advantage of simplicity, it neglects the strategic interaction that would arise when agents cannot precommit to a time path of their control variables. Markov-perfect equilibrium, on the other hand, assumes that players cannot make any commitment of any duration. Reinganum and Stokey (1985) argue that between these two extremes there are possible intermediate formulations; for example, firms can commit to their planned output for the near future, but not for the distant future. To illustrate, they consider a Cournot oligopoly sharing a common stock of an exhaustible resource and having a common terminal date T. The time interval $[0, T]$ is divided into k periods of equal length. It is assumed that at the beginning of each period, firms can make a binding commitment about its output path for that period. If $k = 1$ then firms are able to commit to the whole time path of extraction, that is, the appropriate concept is the open-loop Nash equilibrium. At the other extreme, as k tends to infinity, the length of each period shrinks to zero, and the appropriate concept is the MPNE in continuous time. They then consider the family of games parametrized by the length of the period over which firms can commit to their extraction rates. Assuming isoelastic demand and zero extraction cost, they find that the open-loop Nash equilibrium outcome is socially efficient, despite the common access[12]. On the other hand, as the period of commitment goes to zero, exhaustion takes place at the first instant of time. In between these two extremes, the equilibrium ouput path exhibits downward jumps at the end of each period.

---

[12]This result is in agreement with Weinstein and Zeckhauser (1975) and Stiglitz (1976), who find that monopoly and competition give rise to the same outcome when demand is isoelastic and marginal cost is zero.

3.2.2. *Dynamic games with a non-renewable resource: feedback equilibrium*

Common access without market externalities

In the case of a common property non-renewable resource, MPNE can be easily found if extraction is costless, utility is isoelastic, and each player consumes what he she extracts. Long *et al.* (1999, p. 463), drawing on Long and Shimomura (1998), show that this is true even when players have different discount rates. They prove a theorem that allows comparison between an open-loop Nash equilibrium and an MPNE, showing that the MPNE is equivalent to the OLNE of a similar game in which all agents have a higher discount rate. Another theorem they establish is that an MPNE is equivalent to a OLNE of a game where the stock is subject to a greater rate of depreciation.

Market externalities without common access

It turns out to be very diffult to find analytical solutions for MPNE for a non-renewable resource Cournot oligopoly where each firm has its own reserves (as opposed to the common access case), except for very special cases, such as constant elasticity of demand coupled with zero extraction cost (Eswaran and Lewis, 1985; Reinganum and Stokey, 1985; Benchekroun and Long, 2006), or economic abandonment, such that the resource stocks are not exhausted (Salo and Tahvonen, 2001). Some partial characterizations of closed-loop equilibria or deviation from such equilibria are reported by Polasky (1990) and Benchekroun and Gaudet (2003).

Oligopoly and economic abandonment of resource deposits

In a recent exposition of the state of the oil market, Smith (2009, p. 147) pointed out that *"most of the oil in any given deposit will never be produced, and therefore does not count as proved reserves, because it would be too costly to effect complete recovery."* This indicates that the "exhaustion" of a deposit should be interpreted as an "economic abandonment" of the deposit after the profitable part has been exploited. One way to model economic abandonment is to use the metaphor of a well of water without recharge. When the water surface is close to the ground level, the effort required to raise a bucket of water to the land surface is small. As the water level falls, the required effort per bucket rises. Eventually, the water surface becomes so far away from the ground level that, even though there is still water left in the well, it is no longer worthwhile to draw it. (It would then become more economical to get water from other sources.)

In the case of an industry with heterogeneous firms, using the same metaphor, we may think that each firm owns a well of water, each with a different initial level of water. When the price of "water" is low, only firms whose water level is close to the ground level extract the water. As the price rises, firms with water levels far below the ground level begin to enter the market.

The assumption of economic abandonment renders more tractable the analysis of the MPNE under oligopoly where each firm owns a resource deposit. Explicit strategies have been obtained in analytical form by Salo and Tahvonen (2001) who assume linear demand and marginal extraction cost that increases linearly as the stock level decreases. The implications of this model (based on economic abandonment) are quite different from the model based on eventual exhaustion (as in Lewis and Schmalensee, 1980, and Loury, 1986). The eventual exhaustion model predicts that small firms exhaust their stocks before large firms do, leading possibly to eventual monopolization of the market. In contrast, the economic abandonment model predicts an increase in the number of active firms as the price rises, and the eventual approach to a state where firms with the same cost characteristics have equal market shares, regardless of the initial disadvantage of some of them.

Once the linear demand assumption is dropped, and extraction costs are allowed to be initially independent of the stocks when these are large, Salo and Tahvonen (2001) rely on numerical methods to find a feedback Nash equilibrium.[13] They find that their feedback Nash equilibrium may at first display properties similar to the model of Loury (1986), but eventually it resembles the equilibrium of the economic abandonment model.

3.2.3. *Optimal tariff on exhaustible resources: dynamic games*

If countries that import a non-renewable resource (such as oil) want to extract rent from the foreign suppliers, they can impose a tariff on the imported non-renewable resource. This involves a dynamic game,

---

[13]For an insightful discussion of numerical methods, see Judd (1997). The methods used by Salo and Tahvonen are similar to the approach developed by Haurie *et al.* (1994).

because resource-extracting firms are themselves solving an intertemporal optimization problem (whether they are price takers, or price makers).

Suppose suppliers are price takers, and have perfect foresight (they know the equilibrium price path). Newbery (1976) and Kemp and Long (1980, Essay 16) show that in a dynamic game where a resource-importing country acts as a Stackelberg leader, the optimal time path of the tariff is time-inconsistent, in the sense that at a later date the importing country would want to renege from the previously announced time path of the tariff rate. The intuition for this is as follows. Following Kemp and Long (1980, Essay 16), suppose there are three countries: a resource-exporting country, a strategic resource-importing country (called the home country) that wants to exercise market power, and a passive resource-importing country (called the rest of the world, or ROW).

Assume zero extraction cost and perfectly competitive resource-extracting firms. Then the producer's price must increase over time at a rate equal to the rate of interest, according to Hotelling's Rule. If the strategic resource-importing country announces a time path of a per unit tariff rate, its optimal tariff rate must also increase over time at the rate of interest (so that the domestic consumer's price rises at the rate of interest, i.e., Hotelling's Rule also applies to consumers). Then at some date in the future, the consumers in the home country will stop importing because the price has reached their maximum willingness to pay, that is, the choke price. From that date on, the exporting country is still selling to the rest of the world, as consumers in the ROW are still paying for the resource at a price below the choke price. It will be then in the interest of the strategic resource-importing country to cut the tariff rate so that their consumers can benefit from trade. So, the originally announced tariff path that rises at the rate of interest is time inconsistent. Since producers have perfect foresight, they will recognize from the start this time inconsistency, and they will not believe in the announced tariff path. A recent paper by Keutiben (2009) deals with optimal tariff in a spatial model of trade in exhaustible resources, where both countries own resource stocks. The time-inconsistency of optimal tariff also appears in his open-loop Stackelberg equilibrium.

This inconsistency issue has drawn a lot of attention. Karp (1984) and Karp and Newbery (1991, 1992) impose additional conditions (under various scenarios) to ensure that a time consistent equilibrium path exists when firms are competitive.

Chou and Long (2009) derive Markov-perfect tariff strategies (which overcome the time-inconsistency problem) when the resource-exporting country itself also exercises market power (the case of bilateral monopoly), both for the case where two importing countries form a coalition and for the case where they choose their tariff strategies non-cooperatively. (They restrict attention to price-setting behavior.)

Fujiwara and Long (2009a,b) extend the analysis of Chou and Long to the case where the strategic importing country is a feedback Stackelberg leader, and to the case where it is a feedback Stackelberg follower. They consider both the case of quantity-setting exporter and the case of price-setting exporter.

We explore these issues in greater detail in Chap. 4, which deals with dynamic games in international trade.

3.2.4. *Optimal taxation and principal-agent problems with extractive firms*

The problem of designing a tax-subsidy scheme to induce efficient output can be thought of as a Stackelberg leader–follower problem. Bergstrom *et al.* (1982) show that the government can make a monopolist extractor of a non-renewable resource stock produce at the socially optimal rate by imposing a time path of subsidy. They also show that there is a family of such time paths. (In general the subsidy can be negative, i.e., it is a tax.) Note that in the paper by Bergstrom *et al.* (1981) the per unit subsidy rate is a function of time, that is, it is an open-loop policy.

Karp and Livernois (1992) point out that Bergstrom *et al.* (1981) implicitly assumed that the government can commit to its chosen time path of subsidy. If it cannot, and the monopolist knows this, it will have an incentive to deviate from the extraction path the government wants, because by doing so, he can force the government to change the subsidy path. This means that the subsidy rules proposed by Bergstrom *et al.* are not subgame perfect.

Karp and Livernois show that there is a linear Markov subsidy rule that would be subgame perfect and induce the monopolist to produce the optimal amount (here, linear means that the amount the monopolist has to pay at any time is linear in its extraction rate). In addition, they also show that there is a family of such linear Markov subsidy rules. In the special case of constant extraction cost, Karp and Livernois prove that along the equilibrium path, their linear Markov rule will yield the same tax rate as the open-loop policy if and only if demand is linear or iso-elastic.

Gaudet *et al.* (1995) study the problem of extracting rent from a resource-extracting firm under asymmetric information: the firm has private information about its cost. The government designs a dynamic incentive scheme to maximize social welfare subject to the extractive firm's dynamic incentive constraint. Thus, the government is the Stackelberg leader and the firm is the follower. The authors characterize the nature of the optimal non-linear resource royalty schedule. For related analyses of principal agent problems in a dynamic context, see Gaudet *et al.* (1996, 1998).

3.2.5. *Dynamic games with investment in R&D for a substitute for a non-renewable resource*

Given that a non-renewable resource will eventually be exhausted or abandoned when its extraction cost reaches a prohibitive level, it is natural that countries seek to develop a non-exhaustible substitute, or to explore for new resources. A number of authors analyze single-player problems: Davidson (1978) examines optimal depletion in conjunction with R&D under uncertainty; Dasgupta and Stiglitz analyze the effects of exogenous technological uncertainty on optimal depletion policy; Hung *et al.* (1984) study the optimal timing of transition from an exhaustible resource to an inexhaustible substitute; Quyen (1988) explores the optimal exploration policy under uncertainty. Hoel (1978a, 1979) shows how an exhaustible resource monopoly reacts to the availability of a substitute.

Game theoretic considerations arise when a resource-importing country M seeks to reduce its reliance from a resource-exporting country X by investment in a substitute. Earlier authors, such as Dasgupta *et al.* (1983), Gallini *et al.* (1983), and Olson (1988) focus on the question of the (deterministic) timing of innovation: by spending on R&D, country M can bring closer to the present the date T at which the new technology becomes available. They focus on the case where M is the Stackelberg leader (in the sense that it can commit to its choice of T before country X chooses its output path)[14]. Even with this commitment assumption, the analysis can be quite cumbersome. Dasgupta *et al.* (1983) point to the intriguing possibility that by delaying T a bit, M may induce X to hasten extraction.

---

[14]In contrast, Hung and Quyen (1990) argue that country X is a natural Stackelberg leader. In their model, country M can at any time invent the substitute by paying a fixed lump-sum cost.

Olson (1988) shows that the range of parameter values consistent with this possibility is smaller than previously thought.

The above approach has two shortcomings. First, the assumption of a deterministic innovation date T is not plausible. Second, it is not realistic to assume that a country can commit to a time path of R&D right at the beginning of the game, regardless of subsequent observations of stock levels in the future.

Lewis *et al.* (1986) consider a three-period model in which country M can invest in capacity that begins to produce the substitute one period after the investment. They examine three scenarios: (i) neither country can make a precommitment, (ii) only M can precommit, and (iii) only X can precommit. A number of interesting results emerge. For example, in case (ii) M will overinvest so as to induce a more advantageous extraction path from X.

Harris and Vickers (1995) assume that the invention date is random, and its arrival rate depends only on the intensity of concurrent R&D level. Therefore, the only state variable is the remaining stock of the resource. Their objective is to find a Markov-perfect equilibrium for the game between M and X. Even with one-state variable, the search for a solution is complicated, and Harris and Vickers must rely on an approximate reformulation of the concept of Markov strategy. This is done by allowing X to choose a time path for the resource stock (rather than a decision rule on extraction rate) subject to a set of consistency conditions. They add the simplifying assumption that the resource stock becomes worthless after the invention of the substitute (even though the marginal cost of producing the substitute is positive).[15] An interesting and plausible result is that country M's R&D increases as the resource stock decreases. This fact may induce country X to reduce the rate of decline in the stock and may result in a non-monotone extraction path.

3.2.6. *Common property non-renewable resource and capital accumulation games*

Long and Katayama (2002) consider a differential game model of exploitation of a common property non-renewable resource, where agents

---

[15]This assumption is also made in a special case of Dasgupta and Stiglitz (1981), and yields a model formally equivalent to the model of uncertainty about the date of nationalization by Long (1975).

can also invest in private and productive capital. They show that, in general, there is a phase of capital accumulation before a phase of dissaving. Consumption also reaches a peak before falling. When consumption reaches its peak, net saving is negative. The value function of each player is shown to be separable in the state variables (the resource stock, and the private capital stock). When agents are heterogeneous in terms of productive efficiency, the more productive agents will accumulate more capital, but all agents follow the same consumption rule and the same resource-extraction rule.

In a companion paper, Katayama and Long (2009) use a dynamic game to explore the link between status-seeking and the exploitation of a common-property exhaustible resource, which, together with a stock of man-made capital, are two inputs in the production of a final good. Extraction requires effort; this feature makes the model more general than the standard model of Solow (1974) and Dasgupta and Heal (1979, Chap. 8), where extraction is costless. Katayama and Long (2009) assume that economic agents derive utility not only from absolute consumption, but also from relative consumption, because the latter is a measure of relative status. The authors consider a differential game involving n infinitely lived agents, and compare the MPNE of this game with the outcome under co-operation. They find that the degree of status-consciousness has an important impact on the MPNE. A higher degree of status-consciousness leads to greater excessive consumption, and lower capital accumulation. If extraction is costless, status-consciousness has no impact on the extraction/resource-stock ratio. However, with costly extraction, higher status-consciousness reduces this ratio. This result might seem at first surprising. However, upon reflection, the result is plausible. Since agents want to outdo each other in terms of relative consumption, they find it more efficient (from the individual point of view) to overexploit the common man-made capital stock.

3.3. Related Topics: Recycling, Drug Resistance, and Pesticide Resistance

Gaudet and Long (2003) model a dynamic game between a recyling firm and a primary producer of a mineral product (e.g., aluminum). They focus on the open-loop Nash equilibrium, and compare with the standard result where the primary producer is the Stackelberg leader. Baksi and Long (2009) formulate a dynamic game in discrete time where the primary producer is the Stackelberg leader while consumers themselves are followers who undertake recycling taking into account future prices, and derive some enjoyment from recycling because of social norms.

The resistance of bacteria to drugs has been a concern in the medical profession. Cornes *et al.* (2001) draw attention to the parallel between the decline in effectiveness of antibiotics and insecticide on the one hand, and the exhaustibility of natural resources on the other. They model the use of insecticide by farmers who fail to take into account the effect of their individual applications of doses to the decline in effectiveness of the insecticide. They consider two differential game models: a discrete time model with finite horizon and a continuous time model with infinite horizon. Both models display multiplicity of MPNE that can be ranked according to welfare loss relative to the social optimum. In the continuous time model, in addition to a linear MPNE (which exhausts the effectiveness of the insecticide asymptotically), there are non-linear MPNEs that lead to its exhaustion in finite time.

Chapter 4

DYNAMIC GAMES IN TRADE
AND DEVELOPMENT ECONOMICS

This chapter surveys dynamic game models in international economics and development economics. Section 4.1 is devoted to the effects of exogenously set trade policies (such as voluntary export restraints, quotas, export taxes or subsidies, trade liberalization) on welfare and profits when the market is dominated by international oligopolists that play a dynamic game. Section 4.2 shifts the focus to dynamic games played by two national governments: the classic problem of optimal tariffs and retaliation is examined in the light of the concept of feedback Nash equilibrium, and related equilibrium concepts in dynamic games. Section 4.3 deals with the issue of time-inconsistency of trade policies when a government acts as an open-loop Stackelberg leader, and setting tariffs against competitive foreign suppliers, to exploit terms of trade gain, or to encourage domestic investment. It also deals with ways to achieve time-consistent policies. Sections 4.4 and 4.5 deal with models of trade policies in the presence of an exhaustible resource. Some dynamic game models pertinent to development economics are reviewed in Section 4.6.

4.1. International Oligopoly and Trade Policies

The first generation of dynamic duopoly models of trade began with the pioneering work of Cheng (1987), Soubeyran (1988), Driskill and McCafferty (1989a), and Dockner and Haug (1990, 1991). The sticky-price duopoly model, first proposed by Roos (1925, 1927), and analyzed in detail by Fershtman and Kamien (1987) for both the open-loop equilibrium and the closed-loop equilibrium, proved to be very influential in this stream of contribution to international trade.

The simplest version of the sticky-price model assumes two firms that produce a homogeneous product and that use output as their control variable. The price $p(t)$ at any point of time is given (i.e., it is the state variable), and is not influenced by current outputs. However, if the sum of current outputs, $q_1(t) + q_2(t)$, exceeds the 'notional demand' implied by the current price $p(t)$, then the rate of price change at t is negative. Formally, the notional demand is assumed to be $a - p(t)$ where a is positive constant, and the rate of price change is given by

$$\dot{p}(t) = s[a - p(t) - q_1(t) - q_2(t)] \qquad (4.1)$$

where $s > 0$ is the speed of adjustment. Notice that Eq. (4.1) can be written as

$$p(t) = a - q_1(t) - q_2(t) - \frac{1}{s}\dot{p}(t)$$

which was the functional form proposed by Roos.[1] The cost function is quadratic:

$$C(q_i) = cq_i + \frac{1}{2}q_i^2$$

This linear quadratic model yields linear decision rules that are best replies to each other.[2]

A key feature of the sticky price model of Fershtman and Kamien (1987) is that the equilibrium feedback strategy of each firm has the property that the firm's output at any time t is an increasing function of the currently observed level of the state variable, the price $p(t)$. As a consequence, the industry output in the steady state is higher than the static Nash–Cournot industry output. The intuition behind this result is as follows. A firm's incentive to restrict output to raise the price in the future is weakened by its (correct) perception that the increased price will lead its rival to increase output.

Dockner and Haug (1990) use this model to show that tariff and quota are not equivalent under dynamic duopolistic competition with feedback output strategies. There are two countries, home (H) and foreign (F). The home firm and the foreign firm compete in the market of the home country, H. The foreign firm, firm 2, faces a constant tariff rate τ per unit.

Consider first the open-loop equilibrium formulation, where each firm takes the time path of output of the other firm as given, and chooses its own output path to maximize the present value of its profit stream.[3] Given

---

[1]See Dockner and Haug (1990) for a discussion of this kinematic equation, which can be derived from the assumption that the consumer's current utility depends on past consumption as well as current consumption.

[2]Non-linear decision rules can form feedback equilibria, as noted by Tsutsui and Mino (1990). In this section, we focus on linear decision rules, because they are simpler, and are defined globally.

[3]For an earlier open-loop formulation in industrial organization, see Flaherty (1980).

the tariff τ, the open-loop equilibrium output paths are $\widehat{q}_1(t)$ and $\widehat{q}_2(t)$. Now, suppose the home government replaces the constant tariff rate by a quota path which is exactly equal to $\widehat{q}_2(t)$. Clearly, assuming that the quota is binding, there will be no change in the output paths. This shows that under OLNE there is an equivalence between tariff and quota. Notice that in the limiting case where the rate of interest r goes to zero, the steady-state OLNE outputs and price are identical to the outcomes of the static Cournot model.

Dockner and Haug (1990) then turn to the feedback solution under a constant tariff rate, τ, per unit. This gives rise to two equilibrium feedback output rules, one for each firm, where a firm's output at any time t is conditioned on the currently observed price, $p(t)$. The equilibrium feedback rules exhibit the property that outputs respond positively to increases in the market price. Let q_1^{SS} and q_2^{SS} be the steady-state outputs under this feedback equilibrium. Now, starting at that steady state, replace the tariff rate τ by a constant path of quota equal to q_2^{SS}. Again, assuming that the quota is always binding, would firm 1 maintain its output at q_1^{SS}? The answer is no, because firm 1 now knows that if it reduces its output, and hence causes the future price to rise, firm 2 would not react by raising its output according to its original feedback rule, as it is now subject to a quota. Under the feedback equilibrium with the quota in place, the steady-state price will be higher than that under the feedback equilibrium with a constant tariff rate. Dockner and Haug point out that this result is similar to the static conjectural variations model of Hwang and Mai (1988). This is consistent with the result of Driskill and McCafferty (1989b), who argue that the output strategies of a duopoly under the feedback equilibrium can provide a justification for the static conjectural variations approach.

Dockner and Haug (1990) extend their analysis of feedback equilibrium under a constant tariff and sticky price to the case with n firms. They show that as n goes to infinity, the long-run market prices under tariff and quota are identical in the limit. This result is in accordance to a similar result by Fudenberg and Levine (1988), who show that the open loop and feedback equilibria are approximately the same for dynamic games with many players.[4]

Dockner and Haug (1991) use the same framework to consider the effect of a voluntary export restraint under dynamic duopoly. Recall that

---

[4]Note that the analysis in Fudenberg and Levine (1988) is restricted to the case of a finite horizon.

Harris (1985) shows that under Bertrand competition, a duopoly producing differentiated goods will gain from an export restraint by the foreign firm (and the foreign firm will gain more than the domestic firm) because this turns the foreign firm into a follower and the domestic firm into a leader in the price setting game. Mai and Hwang (1988) show that Harris's result would not apply to a duopoly under static Cournot competition. They demonstrate that an export restraint at the free-trade level will have no effect on price, quantities, and profits. However, they also show that if firms have static negative conjectural variations (i.e., each firm thinks that if it produces more the other firm will produce less) then, starting from a free-trade equilibrium, a voluntary export restraint on the foreign firm will induce the domestic firm to reduce its output to raise the price (it no longer worries that this would provoke the foreign firm to raise its output). This will raise the profit of both firms. This effect is proved in Dockner and Haug (1991) who do not need to assume static negative conjectural variations, because the feedback equilibrium output rule already supplies a similar mechanism in a dynamic setting.

Driskill and McCafferty (1996) use the sticky-price model to reexamine the third market model of static international duopoly (Brander and Spencer, 1985; Eaton and Grossman, 1986, 1988). They show that under quantity competition with sticky prices, the optimal policy for both exporting countries is to tax exports to make the feedback equilibrium of the firms less competitive.[5] The reason is that under the equilibrium feedback output rules (where output responds positively to price) firms tend to be overly competitive: each firm thinks that if it cuts its output, the resulting eventual increase in price will cause the other firm to produce more, undoing its output restriction. Thus, Driskill and McCafferty (1996) provide a justification for the assumption of negative conjectural variations that underlines the export tax result of Eaton and Grossman (1986). As they point out, Eaton and Grossman (1988, p. 603) refer to their use of conjectural variations as "a reduced form way to capture in a simple, static framework the dynamic reactions that make some rivalries more competitive than others".[6] Driskill and Horowitz (1996) extend Driskill and McCafferty (1996) to the case of durable goods.

---

[5]Note that they restrict attention to the limit game, where the rate of interest goes to zero, or the speed of adjustment goes to infinity. Also, they assume that countries are committed to a constant rate of export tax (or subsidy).

[6]Dornbusch (1987, p. 99) and Turnovsky (1986, p. 302) made a similar point.

Fujiwara (2010) extends the Dockner–Haug model to the case where the number of firms in H and F are not the same, and the unit cost of the representative home firm may differ from the foreign one. He shows that under certain conditions, autarky is better than free trade when firms use feedback strategies. This is because the gain in consumer surplus may be dominated by the shifting of profit from home firms to foreign firms. The results of Fujiwara can be extended to allow for horizontal mergers of firms, along the lines of Dockner and Gaunerdorfer (2001) and Benchekroun (2003).

Balboa *et al.* (2007) consider strategic trade policies concerning export rivalry in a third market, when demand is subject to addiction, which they describe as time-non-separable preferences. Home firms and foreign firms use feedback strategies, where the state variable is the degree of addiction which varies with accumulated consumption. The government is committed to a time-invariant rate of export subsidy (or tax). They find the co-existence of the traditional terms of trade motive for an export tax and the profit-shifting motive for an export subsidy. The optimal export tax or subsidy depends on demand and cost parameters. On a similar vein, Yin (2004) considers strategic policies when firms use feedback strategies where the state variable reflects habit formation: the representative consumer in the home market has a demand schedule that depends on both the current price and her/his accumulated consumption.

Fujiwara (2009) uses a differential game to study the welfare effect of trade involving a natural resource and reciprocal dumping. He does not assume sticky prices. The state variable is a stock of renewable resource from which firms extract to obtain inputs. A surprising result is that trade liberalization reduces welfare, at least when the comparison is restricted to the steady-state levels of welfare. This is in contrast to the static reciprocal dumping model, where trade liberalization will increase welfare when both countries are initially near the free trade. A possible explanation lies in the fact that trade involves not only trade costs but also the overexploitation of a common-access resource stock.[7] Fujiwara and Matsueda (2010) consider a model of transboundary stock pollution and its effect on the mode of international competition. Here, the main players are the governments that indirectly decide whether their national firm will be a leader or a follower

---

[7]For some recent models where capital accumulation takes place in an international duopoly, see Calzolari and Lambertini (2006, 2007). Yanase (2005) consider the case where countries suffer from both flow and stock externalities, where the latter may include oligopolistic market interactions.

in international duopolistic rivalry. Yanase (2007) models the choice of
strategies in some dynamic games of environmental policy between two
goverments in a global economy. Taxes are compared to quotas.

An entirely different type of dynamic international duopoly is
considered by Fujiwara and Long (2010a). They model a dynamic contest
between a home firm and a foreign firm that compete for a government
procurement contract at each point of time. The state variable is their
relative stock of influence (or goodwill), which they need to have a chance
to win a contest. They invest in their stock of influence by cultivating their
relationship with the bureaucrats who collectively make the decision about
which firm will win a particular contract. Similar to advertising models,
where a firm's goodwill is a summary measure of the favorable attitude of
its potential customers toward its product, in this contest model, a firm's
relative goodwill is a measure of the favorable disposition of the bureaucrats
who choose the winner. Denote by $w(t)$ the domestic firm's relative stock
of influence, measured on a scale such that its smallest possible value is
zero, and its greatest possible value is unity. Thus, $w^*(t) \equiv 1 - w(t)$ is the
foreign firm's relative stock of goodwill. The probability that the domestic
firm wins the period t procurement contract is $p(w(t))$, where $p(w)$ is an
increasing function, with $p(0) = 0$ and $p(1) = 1$. Then, $1 - p(w(t))$ is
the probability that the foreign firm wins. Assume $p(w) = w$ identically.
At any point of time, the relative stock of goodwill is given. Over time,
goodwill can be cultivated. The authors model the evolution of $w(t)$ as
follows. Assume that $\dot{w}(t)$ depends on the effective amount of money the
domestic firm spends on lobbying, $s(t)$, in relation to its rival's effective
spending, $\beta s^*(t)$, and its stock $w(t)$. Here, $\beta \in [0,1]$ is a parameter that
converts the foreign firm's nominal lobbying expenditure $s^*(t)$ into effective
spending. For example, if $\beta = 0.75$, then only 75 cents out of each dollar
that the foreign firm spends on lobbying are effective. Thus, β is inversely
related to the degree of bias against the foreign firm. If $\beta = 1$, there is no
bias. The closer β is to zero, the greater the bias. Fujiwara and Long (2010)
assume

$$\dot{w} = \alpha \left[\ln \left(\frac{s}{w} \right) - \ln \left(\frac{\beta s^*}{1-w} \right) \right] (1-w)w, \quad w(0) \in (0,1) \text{ given} \qquad (4.2)$$

where $\alpha > 0$ is a parameter of the speed of adjustment of w. The term
$(1-w)w$ on the RHS ensures that w can never become negative, and it
can never exceed 1. The term inside the square brackets indicates that the
domestic firm's stock of goodwill increases if and only if the ratio of its

lobbying expenditure to its relative goodwill, s/w exceeds the foreign ratio $\beta s^*/(1-w)$. By definition, the domestic government treats the foreign firm more equally as β increases toward 1. Trade liberalization in procurements is modeled as an exogenous increase in β.

Let V denote the domestic firm's gross profit if it wins the contest. The domestic government levies a profit tax rate, τ, on each firm. Taking τ as given, the domestic and foreign firms non-cooperatively choose their time paths of lobbying efforts to maximize their discounted streams of expected net profit. The underlying environment is formally described by the following dynamic game:

$$\max_{s} \int_0^\infty e^{-rt}[w(1-\tau)V - s]dt$$

$$\max_{s^*} \int_0^\infty e^{-rt}[(1-w)(1-\tau)V^* - s^*]dt$$

subject to the transition equation (4.2), and where $w \in [0,1]$ is the probability with which the domestic firm wins the contest. Here, $r > 0$ is a constant rate of discount. The authors focus on the open-loop Nash equilibrium. They find that trade liberalization, in the form of a reduction in bias against the foreign firm, improves both domestic and global welfare if (i) either the foreign firm's profit is sufficiently large or (ii) the initial degree of home bias is sufficiently small. If the initial home bias is large, a small reduction in the bias may reduce welfare.

4.2. Feedback Trade Policies under Bilateral Monopoly

This section shifts the focus from dynamic games among firms to dynamic games between governments that choose the level of their trade policy instruments to maximize national welfare, taking into account the state dynamics. The most immediate example is the setting of optimal tariffs in a dynamic context.

The theory of optimal tariffs is mostly known in the static framework. There was some discussion of tariffs and retaliation in Johnson (1954) but the story is based on myopic optimization in each period. There are models of repeated games of tariff setting, but repeated game models are not truly dynamic because the state of the system remains unchanged over time.

Kemp *et al.* (2001) consider a differential game model of trade war involving a durable good (say, cars) and a perishable good (say, food). In that model, the representative consumer in country i has the utility function

$$U = D_i(t) + \ln C_i(t)$$

where $D_i(t)$ is the service flow from the durable good stock $K_i(t)$ that she/he owns. In particular, assume $D_i(t) = sK_i(t)$ where s is a positive parameter. The stock is subject to depreciation, and can be replenished by additional purchases of the newly produced durable good. Suppose the home country imports food and exports cars. It imposes an *ad valorem* tariff rate $\tau(t)$ on food. The foreign country imposes an *ad valorem* tariff rate $\varepsilon(t)$ on cars. By Lerner's symmetry, this is equivalent to an export tax on food. The game is played between the two governments, each seeking a tariff rate rule that would maximize the present value of the life-time welfare of its representative consumer. It turns out that in the feedback Nash equilibrium *ad valorem* tariff rules are independent of the state variables $K_1(t)$ and $K_2(t)$. This is an outcome of the assumption that utility is linear in the state variables. The authors show that tariff wars can be beneficial to one country and harmful to the other, if their endowments are sufficiently different from each other.

Chou and Long (2009) use differential game techniques to analyze the impact of bilateral monopoly on the world trade in exhaustible resources, and on the welfare level of importing and exporting nations. Building on the model of Liski and Tahvonen (2004), they assume that the resource stock will eventually be abandoned because of rising extraction costs. The resource-importing nations choose non-cooperatively their feedback tariff rules which condition the tariff on the size of the remaining resource stock, while the resource-exporting country chooses a feedback pricing rule that determines the monopoly's current price as a function of its remaining stock. The authors specialize in the linear quadratic case, and compare the scenario where importing nations (of different sizes) do not coordinate their tariff policies with the scenario where they set a common tariff. Of particular interest is the impact of asymmetry on welfare. As the asymmetry in market sizes of the two importing countries increases, the exporting country's welfare decreases in both the free trade case and the case of trade war with non-cooperative tariff setting, while the sum of gains by importing countries increases. When there is only one exporting country and one importing country, the feedback Nash equilibrium can be calculated analytically.

Fujiwara and Long (2010b) compute the feedback Stackelberg equilibrium of the two-country version of the model of Chou and Long

(2009), first for the case where the price-setting exporting country is the leader, and then for the case where it is the follower (i.e., it chooses its producer-price rule after learning the feedback tariff rule used by the importing country).

In Fujiwara and Long (2009), the exporting firm uses a quantity setting rule: output at any time is a function of the remaining stock. They assume there are two importing countries, one of which is committed to free trade, while the other country (called Home) uses a tariff setting rule: the per unit tariff rate at any time depends on the remaining stock. They found that under either the leadership of Home or that of the exporter, both parties are better off compared with the feedback Nash equilibrium. In particular, both parties prefer Home to be the leader.

4.3. Time-inconsistency of Open-loop Trade Policies

It has been long recognized that in dynamic games involving a Stackelberg leader and a follower (or many followers), the problem of time-inconsistency typically arises.[8] If the leader announces a time path of its future actions (e.g., future tariff rates) to induce the followers to take actions now (e.g., specific investment) that would benefit it in the future, it is likely that at some future time the leader would not want to honor its commitment. In the context of trade involving an exhaustible resource, Newbery (1976) and Kemp and Long (1980d) demonstrate that the optimal tariff is generally time-inconsistent. Suppliers of the resource make their current output decision on the basis of the future time path of the tariff rate which the importing country announces at the beginning of the game to induce a favorable supply response. However, the importing country will at some stage in the future have no incentive to stick to the initially announced time path of the tariff rate.

Eaton and Grossman (1985) and Staiger and Tabellini (1987) also discuss the time-inconsistency of tariffs. In both the papers, tariffs are used as second-best tools, to redistribute income when the insurance markets are imperfect. While the numerical examples of Eaton and Grossman show that the optimal and time-consistent tariffs are similar, Staiger and Tabellini report that their time-consistent solutions are more protectionist. Lapan

---

[8]See Simaan and Cruz (1973b, p. 619), Kydland (1975), Kydland and Prescott (1977), Hiller and Malcomson (1984).

(1988) points out that for any large country, if production decisions occur before consumption decisions, there is an *ex post* incentive to increase tariffs above the *ex ante* optimal level. Using a model with a large country and a collection of small identical foreign countries, Lapan (1988) shows that *all countries are worse off* if the large country cannot precommit to its *ex ante* optimal tariff. Lapan's arguments show that the standard textbook formula that the optimal tariff is equal to the reciprocal of the foreign price elasticity of supply ignores an important question: the distinction between short-run and long-run elasticities. This distinction is taken up by Karp (1987), who considers a truly dynamic model.

Karp (1987) points out that the equilibrium tariff path depends both on the information available to the sellers in the exporting country (foreign, or F) and on the type of strategy available to the importing country (home, or H). To illustrate, Karp assumes that foreign firms are large in number, and face costs of adjusting output. Assume that the production function of the representative firm is $y = K$ where K is its capital stock and y is its output. Let I denote its rate of gross investment. The representative foreign firm takes the price path as given, and chooses a time path of investment to maximize its intertemporal profit:

$$\int_0^\infty e^{-rt} \left[p(t)K(t) - vI(t) - \frac{cI(t)^2}{2} \right] dt$$

subject to

$$\dot{K}(t) = I(t) - \delta K(t), \quad K(0) \text{ given} \tag{4.3}$$

Here, v is the unit cost of investment and c is the parameter of the adjustment cost function. Let ψ be the shadow price of the capital stock. The foreign firm's optimality conditions are

$$I = \frac{\psi - v}{c} \tag{4.4}$$

$$\dot{\psi} = (r + \delta)\psi - p \tag{4.5}$$

and the transversality condition is

$$\lim_{t \to \infty} e^{-rt}\psi(t)K(t) = 0 \tag{4.6}$$

Given the time path $p(.)$, the four Eqs. (4.3)–(4.6) yield the time paths of K and I. Notice that the initial value of the shadow price, ψ_0, depends on the entire time path of p.

Assume that in country F the demand for the product is $D_F(p)$. Then at time t, the excess supply faced by the home country is $X(t, p(t)) \equiv K(t) - D_F(p(t))$. The short-run price elasticity of export supply is

$$\varepsilon_{\text{SR}} = \frac{p(t)}{X}\frac{\partial X}{\partial p(t)} = \left[\frac{p}{K(t) - D_F(p(t))}\right]D_F'(p(t))$$

In the long run, $K(t) \to K_\infty$ where K_∞ depends on the limiting behavior of the time path of p. If $p(t) \to p_\infty$, then

$$K \to K_\infty = \frac{p_\infty}{(r + \delta)\delta c} - \frac{v}{\delta c}$$

Thus, the long-run price elasticity of export supply is

$$\varepsilon_{\text{LR}} = \frac{p_\infty}{X}\left[\frac{\partial X}{\partial p_\infty} + \frac{\partial X}{\partial K_\infty}\frac{\partial K_\infty}{\partial p_\infty}\right]$$

The welfare of the home country is

$$\int_0^\infty e^{-rt}\{U[K - D_F(p)] - p[K - D_F(p)]\}dt$$

where $U(.)$ is the utility of consuming the imported good. The home government, by choosing the time path of p, can influence the time path of (ψ, K, I). Karp proposes the transformation

$$z(t) \equiv \dot{K}(t) = I(t) - \delta(t)$$

Then, the home country faces the system of equations

$$\dot{z} = rz + (r + \delta)\delta K + \frac{(r + \delta)v - p}{c}, \quad z(0) \text{ free}$$
$$\dot{K} = z, \quad K(0) \text{ given}$$

At time $t = 0$, the home country can "choose" $z(0)$ by announcing the time path of p. Suppose its optimal choice is $z^*(0)$. This implies a particular time path of z. Karp (1987) uses this formulation to show that, in the absence of commitment, time-inconsistency will occur. At any time $t_1 > 0$, when a new policymaker takes over, he/she will see that the foreign firms have sunk their investment, and will deduce (correctly) that it is in the interest of the home country to choose a new initial value $z(t_1)$ which differs from the one implied by $z^*(0)$.

In general time-inconsistency will occur, unless the leader can enforce a "command optimum" (i.e., the follower's behavior is no longer a constraint).[9]

Other discussions of the problem of time-inconsistency in trade policies include Tornell (1991) and Miyagiwa and Ohno (1999). Leahy and Neary (1999) use a two-period model to discuss, among other scenarios, the commitment problem on a period-by-period basis in the context of infant industry promotion. Matsuyama (1990) characterizes the perfect equilibrium for a model where a domestic firm and the government play a repeated bargaining game: the firm asks for temporary protection to develop a cost-reducing investment, while the government wants to liberalize trade to maximize social welfare. There is no state variable in the model, so the game is not a dynamic game in our view.

Miravete (2003) introduces dynamics into Matsuyama's model by allowing the firm's level of marginal cost to fall over time as a result of accumulated output (due to learning by doing). The author characterizes the time-consistent tariff policy in the case where the protected firm (the domestic monopolist) also optimizes intertemporally. The state variable is the level of the monopolist's marginal cost. Even though the game is linear quadratic, explicit solutions of feedback strategies are not available.[10] Miravete (2003) finds that under certain assumptions there is a unique MPNE, and it leads to the future liberalization of trade.[11]

4.4. Optimal Export Tax by a Resource-Exporting Country

Kemp and Long (1979) model bilateral trade between a resource-rich and a resource-poor economy. They assume that the resource-rich country exports its extracted resource which is used as an *essential input* in the resource-poor economy.[12] Under the assumption of costless extraction, they characterize a competive equilibrium under the requirement that trade

---

[9]Hiller and Malcomson (1984) construct an example where by having an additional tax instrument the government can achieve the command optimum. The use of history-dependent control rules may enable a command optimum, see Basar and Selbuz (1979), Papavassilopoulos and Cruz (1980).

[10]Each of the Ricatti equations that the linear strategies of this model must satisfy corresponds to a hyperbola (McLenaghan and Levy, 1996). Miravete uses stability conditions to eliminate some candidate solutions.

[11]Lockwood (1996) shows that there are sufficient conditions for uniqueness of linear Markov-perfect equilibrium in infinite horizon affine-quadratic differential game.

[12]Note that the "choke price" of an essential input is infinite.

is balanced at each point of time (and there is no international capital market).[13] They next consider the case where the resource-rich country knows that it can influence the terms of trade by interfering with its export flow. In general, the time path of production will change. The resource-rich country's optimal extraction path can be achieved by control and command, or by a suitable time path of export tax. Assume the government of the resource-rich country announces its entire time path of *ad valorem* export tax rate $\tau_1(t)$, that is, it acts as an open-loop Stackelberg leader. Kemp and Long show that if $m_1^*(t)$ denotes the optimal extraction path, then the open-loop Stackelberg leader's optimal *ad valorem* export tax rate $\tau_1^*(t)$ satisfies the following condition:

$$1 + \tau_1^*(t) = \frac{1}{\beta[1 + \delta(m_1^*(t))]}$$

where $\delta(m)$ is the elasticity of the marginal productivity of the resource input, and β is an arbitrary positive constant. This shows that the producer's price $p^1(t)$ of the exhaustible resource (in terms of the consumption good) in the resource-exporting country is related to the world price (the terms of trade) $p^*(t)$ by

$$p^1(t) = p^*(t)\beta[1 + \delta(m_1^*(t))]$$

Since β is an arbitrary positive constant, this result shows that the only the rate of change in the tax rate matters, not its level. This is because if the time path $p^1(t)$ is multiplied by a constant, extracting firms will not change their supply behavior, and it is a matter of indifference to consumers whether they receive their income as dividends or tax handouts.

In the special case where the production function of the resource-rich country is Cobb–Douglas, such that the input demand is isoelastic, the resource-rich country has no market power. This is consistent with the result of Weinstein and Zeckhauser (1975) and Stiglitz (1976) that under zero

---

[13]Chiarella (1980) generalizes Kemp and Long (1979) by allowing for capital accumulation in the resource-poor economy. He assumes that all factors are traded on international markets and each country chooses its time path of its quantity variables, taking as given the factor-price time path. The equilibrium price path is the one that equates demand and supply. It is as if there is an auctioneer in the backgound. This "differential game" therefore results in a dynamic competitive equilibrium (a continuous-time, infinite horizon version of the familar Arrow–Debreu model). Assuming a constant rate of technical progress, Chiarella shows that in the absence of a capital market, the trading equilibrium generally violates the Solow–Stiglitz efficiency condition along the adjustment path to the steady state. With the capital market, the outcome is efficient.

extraction cost and constant elasticity of demand, the exhaustible resource monopoly has no market power.

What happens if the resource-poor country exercises its market power while the resource-rich country is passive? In this case, the resource-poor country can simply offer a price of zero forever. The situation would change if the resource-rich country also exercises its power. This is the case of bilateral monopoly, and if the equilibrium concept is the feedback Nash equilibrium, the solution will depend on whether the exporting monopolist is committed to a price decision rule or a quantity decision rule.[14]

4.5. Optimal Tariff by Resource-Importing Countries

If a government of a country imposes a tariff on its imports of natural resources, it is in fact playing a dynamic game against suppliers of the good, each solving its own optimal control problem to determine its output strategy (or price strategy). The analysis of interaction between the importing country and the sellers of a non-renewable resource can be quite complicated, as we shall see later.

4.5.1. *Optimal tariff under rational expectations, when sellers are perfectly competitive*

Bergstrom (1982) considers tariff policies by a group of resource-importing countries facing competitive suppliers (from a resource-exporting country). He then asks what *constant ad valorem* tariff rate each importing country would independently impose in a Nash equilibrium where each importing country's objective is to maximize the present value of the stream of net benefits (the sum of consumer's surplus and tariff revenue). Bergstrom shows that importing countries can extract substantial rents from resource owners. Bergstrom points out that relaxing the simplifying assumption of a *constant ad valorem* tariff would result in a much more difficult task, both conceptually and computationally, though in the special case where the demand function exhibits constant elasticity and extraction is costless there is a Nash equilibrium where each country would choose a constant *ad valorem* tariff.[15]

---

[14]For other strategic aspects of trade involving an exhaustible resource, see Hillman and Long (1983, 1985).

[15]Brander and Djajic (1983) also assume a constant *ad valorem* tariff. In their model, the exporting country increases the domestic demand of the resource to raise the world price.

Time-inconsistency under open-loop optimal tariff

Newbery (1976) and Kemp and Long (1980d) explore the optimal tariff on an exhaustible resource without imposing the assumption of a constant tariff path. They find that the open-loop optimal tariff path is indeed time-inconsistent. The intuition for this result is best illustrated by using the three-country formulation of the oil market proposed by Kemp and Long (1980). Their world consists of a strategic oil importing country (called the home country, H), a passive oil-importing country (that adopts free trade), and an oil-exporting country. Consumers in both importing countries will buy oil only if their domestic price is less than a given number a, called the choke price. The oil-exporting country has a large number of resource-extracting firms that are price takers. With zero extraction cost, along an equilibrium path, the producers' price must rise exponentially at the rate of interest. The strategic importing country takes this dynamics of the producers' price as a constraint, and chooses a time path of specific tariff rates to maximize its welfare. Clearly for any total amount of oil import over the importing phase, the strategic oil-importing country must allocate oil to its consumers efficiently. This requires that their marginal utility of oil consumption rise at the rate of interest. It follows that the optimal specific tariff rate must also rise at the rate of interest. Since the world price is lower than the domestic price in H by the tariff amount, by the time H's domestic price reaches the choke price a, trade in oil is still taking place in the rest of the world. By that time the government of H would find it advantageous to allow imports, thus dropping its initial commitment to the initially announced path of tariff rates. (By continuity, even before that time, the government of H would have an incentive to allow more imports than previously planned.)

The results of Newbery (1976) and Kemp and Long (1980d) indicate that a time-consistent tariff would be found only if the open-loop formulation is dropped. This issue is taken up by Karp (1984), Maskin and Newbery (1990), and Karp and Newbery (1991, 1992).

Time-consistent tariff and the disadvantage of having market power

In Karp and Newbery (1992), it is assumed that the perfectly competitive sellers collectively behave as if they followed a Markovian decision rule: aggregate extraction at time t, denoted by $y(t)$, is a function of the aggregate stock of oil that remains, $S(t)$. In country i, the representative consumer has the demand function $q_i = b_i(a - p_i)$ where a is called the choke price, p_i is the tariff-inclusive price, that is, $p_i = p + \tau_i$, and p is the world price. The instantaneous welfare of importing country i, defined as the sum of consumer's surplus and tariff revenue, is given by

$$W_i = \frac{b_i}{2}(a - p - \tau_i)(a - p + \tau_i) \qquad (4.7)$$

Assume that there are only two importing countries and that the price $p(t)$ is determined by the market-clearing condition, $q_1(t) + q_2(t) = y(t)$. Then, normalizing such that $b_1 + b_2 = 1$, the market-clearing price is given by

$$p(t) = a - y(t) - b_1\tau_1(t) - b_2\tau_2(t)$$

All players move simultaneously. Suppose that importing country 1 believes that country 2 uses a Markovian tariff rule $\tau_2 = g_2(S)$ and that the suppliers collectively follow the rule $y = Y(S)$. Then, its instantaneous welfare is

$$W_1(\tau_1, S) = \frac{b_1}{2}[Y(S) + b_2g_2(S) - (1 - b_1)\tau_1][Y(S) + b_2g_2(S) + (1 + b_1)\tau_1]$$

Given $Y(S)$ and $g_2(S)$, country 1 chooses a path of tariff rate $\tau_1(.)$ to maximize the integral of the discounted stream of its instantaneous welfare $W_1(t)$. Suppose all countries use the same rate of discount $r > 0$. Let $V_1(S)$ be the value function for country 1. Since $\dot{S} = -Y(S)$, its Hamilton–Jacobi–Bellman (HJB) equation is

$$rV_1(S) = \max_{\tau_1}[W_1(\tau_1, S) - V_1'(S)Y(S)]$$

Since country 1 takes the supply rule $Y(S)$ as given, independent of its tariff rate, its HJB equation gives the simple first-order condition:

$$b_1[Y(S) + b_1\tau_1 + b_2g_2(S)] = \tau_1$$

that is, $\tau_1 = b_1(a - p)$. Suppose country 2 acts similarly. Then

$$b_2[Y(S) + b_2\tau_2 + b_1g_1(S)] = \tau_2$$

that is, $\tau_2 = b_2(a - p)$. Hence, we obtain

$$\frac{\tau_1}{\tau_2} = \frac{b_1}{b_2}$$

In equilibrium, the strategies satisfy

$$g_i(S) = b_i(1 - b_1^2 - b_2^2)Y(S)$$

The equilibrium price can be expressed as a function of S:

$$p(S) = a - [1 - (b_1^2 + b_2^2) + (b_1^2 + b_2^2)^2]Y(S) \equiv a - \mu Y(S)$$

Now, we must determine the function $Y(S)$. Recall that the Hotelling Rule tells us that the price must rise at the rate of interest in the competitive market for an exhaustible resource with zero extraction cost. Then

$$r = \frac{\dot{p}}{p} = \frac{p'(S)}{p(S)}\dot{S}$$

$$r = \frac{\mu Y'(S)Y(S)}{a - \mu Y(S)}$$

This equation is a first-order differential equation. Let us impose the boundary condition that $Y(0) = 0$, that is, if the remaining stock is zero, then the extraction is zero. We can then compute the equilibrium supply rule $Y(S)$ numerically. Notice that, since $Y(0) = 0$, the above equation implies that

$$\lim_{S \to 0} Y'(S) = \infty$$

Karp and Newbery (1992) compute the equilibrium price path and welfare levels of each player. Comparison with the free-trade case indicates that the importing countries can be worse off compared with free trade, but unless commitment to free trade is possible and credible, free trade is not an equilibrium of the game where suppliers believe that buyers have market power.

Maskin and Newbery (1990) provide another example of disadvantageous market power. Their model is in discrete time, and the dominant importing country is the first mover in each period. This is in contrast to Karp and Newbery (1992) where players move simultaneously at each point of time.

Karp and Newbery (1991) compare two models of the exercise of market power by buyers in the face of perfectly competitive suppliers of oil. The first model, called the exporters move first (EMF) model, assumes that the sellers move first in each period (and in the limit, at every instant). In the limit, as the length of each period shrinks to zero, the equilibrium of this model is the same as that of Karp and Newbery (1992). The second model, called the importers move first (IMF) model, assumes that in each period, the importers move first. Using Eq. (4.7), and noting that $q_i = b_i(a - p - \tau_i)$, we can express the welfare of importing country i as

$$W_i = q_i \left(a - p - \frac{q_i}{2b_i} \right)$$

In the limiting case when the length of each period tends to zero, we obtain a differential game in continuous time. The HJB equation for the importing country i is then

$$rV_i(S) = \max_q \left[q_i \left(a - \widehat{p}(S) - \frac{q_i}{2b_i} \right) - V_i'(S)(D_{-i}(S) - q_i) \right]$$

where it is assumed that the country takes as given the Markovian "equilibrium price function", $\widehat{p}(S)$, and the equilibrium consumption strategy of the other importing countries, denoted by $D_{-i}(S)$. Assume $b_i = 1/n$. Then the HJB equation, after maximization with respect to q_i, becomes

$$rV_i(S) = \frac{(a - \widehat{p}(S) - (2n - 1)V_i'(S))}{2n}(a - \widehat{p}(S) - V_i'(S)) \qquad (4.8)$$

On the other hand, under zero extraction cost, the price must rise at the rate of interest. The Hotelling Rule can be expressed as

$$V_i'(S) = \frac{r\widehat{p}(S)}{\widehat{p}'(S)} + (a - \widehat{p}(S)) \qquad (4.9)$$

Equations (4.8) and (4.9) constitute a system of two first-order differential equations, with the boundary conditions $\widehat{p}(0) = a$ and $V_i(0) = 0$. Numerical solutions indicate that, for the representative importing country, the ratio of its welfare under EMF to that under IMF depends on the initial stock S_0. For low values of S_0, being the first mover is a disadvantage to importers. However, for very large values of S_0, importers do better under IMF than under EMF.

Time-consistent tariff under economic abandonment

The above models of time-consistent tariff share the common assumption that the resource is homogeneous and will be eventually exhausted. An alternative assumption is that the extraction cost rises as the stock dwindles. When the extraction cost becomes as high as the choke price the resource will be abandoned. Would the problem of the time-inconsistency of the open-loop optimal tariff arise under this scenario? If yes, what would be a reasonable restriction on equilibrium strategies to ensure time-consistency? Karp (1984) explores these issues in a two-country model: a resource-importing country and a resource-exporting country that consists of competitive extractive firms. Below is a variant of Karp's model.

Assume that the inverse demand function is $P = a - Q$ where Q is the quantity demanded, P is the price faced by consumers of the importing country (inclusive of tariff), and a is the choke price. Consider a representative resource-extracting firm with an initial stock of resource, Z_0. We denote its remaining stock at any time $t > 0$ by $Z(t)$. Let $y(t)$ denote its extraction rate, such that $\dot{Z}(t) = -y(t)$. The cost of extracting $y(t)$ is $c(Z(t))y(t)$, where $c(Z_0) = c_0 \geq 0$ and $c'(Z(t)) < 0$ indicating that as the remaining stock gets smaller, the extraction costs rise.

Let $\tau(t)$ be the per unit tariff. The firm takes the time path of net price it faces, $P(t) - \tau(t)$, as given. Let $\widehat{Z}$ be the stock level such that $c(\widehat{Z}) = a$. Clearly, if $\tau(t) \geq 0$ then $P(t) - \tau(t) \leq a$, and the firm will abandon its deposit as soon as (or before) Z falls to $\widehat{Z}$. Let $\mu(t) \geq 0$ be the firm's co-state variable for the stock $Z(t)$. Its extraction path $y(t)$ must satisfy the first-order conditions

$$e^{-rt}[P(t) - \tau(t) - c(Z(t))] - \mu(t) = 0$$

and

$$\dot{\mu}(t) = e^{-rt}c'(Z(t))y(t) < 0 \qquad (4.10)$$

Let $u(Q)$ be the buyer's utility of consuming Q, where $u'(Q) = a - Q$. The importing country pays the seller $P(t) - \tau(t)$ for each unit of good it imports. Its objective is to maximize

$$J_M = \int_0^T e^{-rt}[u(Q) - (P - \tau)Q]\,dt$$

The time T is endogenously determined. In equilibrium, $Q = y$ and

$$P - \tau = c(Z) + e^{rt}\mu$$

After substituting for Q and $P - \tau$, we get an alternative expression for the importing country's objective function:

$$J_M = \int_0^T \left\{ -\mu y + e^{-rt}\left[u(y) - c(Z)y\right] \right\} dt \qquad (4.11)$$

The term $e^{-rt}[u(y) - c(Z)y]$ represents the world's net gain from extracting y. It follows that μy can be interpreted as that part of the world's net gain that the supplier captures. Integrating Eq. (4.10) to get

$$\mu(T) - \mu(t) = \int_t^T e^{-r\theta} c'(Z(\theta)) y(\theta) \, d\theta$$

Using this to substitute for $\mu(t)$, we obtain

$$J_M = \int_0^T \{-\mu(T)y + e^{-rt}[u(y) - c(Z)y]\} \, dt$$

$$+ \int_0^T y(t) \left[\int_t^T e^{-r\theta} c'(Z(\theta)) y(\theta) d\theta \right] dt$$

The second integral is of the form $\int_0^T S'(t)G(t)dt$ where $S(t) \equiv \int_0^t y(\theta)d\theta = Z_0 - Z(t)$, and $G'(t) = -e^{-rt}c'(Z(t))y(t)$. Note that $G(T) = 0$ and $S(0) = 0$. Integration by parts yields

$$\int_0^T S'(t)G(t)dt = S(T)G(T) - S(0)G(0) - \int_0^T S(t)G'(t)dt$$

$$= \int_0^T [Z_0 - Z(t)]y(t)c'(Z(t))e^{-rt}dt$$

Thus, the importing country's objective function becomes

$$J_M = \int_0^T e^{-rt}[u(y) - c(Z)y + (Z_0 - Z)yc'(Z)]dt - \mu(T) \int_0^T y(t)dt$$

(4.12)

Since $\mu(t) \geq 0$, to maximize this integral the importing country should set $\mu(T) = 0$, and choose the time path of y to maximize the first integral in (4.12), subject to $\dot{Z} = -y$, $Z(0) = Z_0$, and $Z(T) \geq 0$. The term Z_0 in the objective function indicates that if $c'(.) \neq 0$, the importing country's optimal plan formulated at time 0 will be time-inconsistent. This is because if it can replan at some future time $t_1 \in (0, T)$ then its objective function at time t_1 will be

$$\int_{t_1}^T e^{-rt}[u(y) - c(Z)y + (Z(t_1) - Z)yc'(Z)] \, dt$$

since Z_0 is no longer relevant. This modified objective function will result in a different time path for y from t_1 to T. This is Karp's proof that if the extraction cost is stock dependent the open-loop optimal tariff is time-inconsistent even in a two-country model where demand comes only from the importing country.

To recapitulate, we have tried to solve the following Stackelberg leadership problem: maximize the integral

$$J_M = \int_0^T \{-\mu y + e^{-rt}[u(y) - c(Z)y]\}dt \qquad (4.13)$$

subject to the following constraints

$$\dot{Z} = -y, \qquad Z(0) = Z_0, \qquad Z(t) \geq 0 \qquad (4.14)$$

$$\dot{\mu} = e^{-rt}c'(Z)y, \mu(0) \text{ free}, \mu(t) \geq 0. \qquad (4.15)$$

and we have found that the solution is time-inconsistent.

To find a time-consistent tariff policy, Karp (1984) proposes that one solves instead the modified problem of maximizing

$$\widehat{J}_M = \int_0^T e^{-rt}[u(y) - c(Z)y]\,dt \qquad (4.16)$$

(i.e., without the term $-\mu y$) and subject only to constraint (4.14), that is, one simply ignores the constraint (4.15). After solving this modified problem, and letting an asterisk indicate the values of variables in this solution, one computes the welfare of the importing country under the consistent tariff as follows:

$$J_M^C = \widehat{J}_M - \int_0^T \mu^*(t)y^*(t)\,dt$$

where the superscript C indicates that the time-consistency requirement is imposed. As Karp shows, J_M^C is lower than J_M (obtained under the time-inconsistent open-loop Stackelberg leadership).

Under Karp's proposed solution method in the case of competitive firms, the importing country is behaving as if it would want to maximize world welfare. If the seller is a monopolist then, for this method to work, it would be necessary to assume that the monopolist behaves as if he did not take into account the feedback decision rule $\tau(t) = \phi(Z(t))$.

4.5.2. *Optimal tariff when the seller is a monopolist*

When the seller is a monopolist, the importing country must realize that it is playing a dynamic game of bilateral monopoly. Some papers deal with this issue in a context that involves pollution externalities. We refer the reader to Chap. 2 (see in particular the section on carbon taxes). The pure case (without carbon taxes) of bilateral monopoly is considered by Rubio (2005),

Chou and Long (2009), and Fujiwara and Long (2009c, 2010b). Rubio and Escriche (2001) compare the MPNE with the stage wise Stackelberg equilibrium when the seller is the (stage wise) Stackelberg leader and find that the two equilibria are identical.[16] Fujiwara and Long (2009b) use a different concept of Stackelberg equibrium, and find that when the seller is the (non-stage wise) Stackelberg leader, he/she earns higher profit than under the Nash equilibrium.

4.6. Foreign Aid and Capital Flight from Poor Countries to Rich Countries

Let us turn to some applications of dynamic games to development economics. We first look at some models of capital flows from rich countries to poor countries, in the form of foreign aid. Afterward, we look at some models where capital flows from poor economies to rich economies because of corruption and imperfect property rights.

4.6.1. *Foreign aid as a dynamic game*

There is a wealth of literature on static models of foreign aid. The classic transfer problem (see, e.g., Samuelson, 1952) may be thought of as an early analysis of some aspects of foreign aid. Kemp (1984) notes that international transfers, when conceived as voluntary contributions to a public good, are independent of small changes in the distribution of wealth among donor countries. Kemp and Shimomura (2002) remark that the theory of voluntary and unrequited international transfer rests on two incompatible assumptions: (i) each country is indifferent to the wellbeing of other countries, and (ii) voluntary unrequited international transfers do take place. They therefore propose a more satisfactory model that would allow for the possibility that the wellbeing of each country is influenced by the wellbeing of other countries, and characterize the optimal foreign aid from the point of view of the donor. Kemp *et al.* (1992) formulate a model of dynamic foreign aid with capital accumulation, in the spirit of an infinite horizon principal agent model (but without moral hazard).

Kemp and Long (2009) offer two models of foreign aid where there are several donor countries that do not coordinate their aid strategies. In the first model, donor countries continually feel the warm glow from the

---

[16]For an explanation of the concept of stagewise Stackelberg leadership, see Chap. 2.

act of giving. The feedback Nash equilibrium aid strategies turn out to be linear strategy $A_i = \alpha^* X$, where $X(t)$ is the capital stock of the recipient country and $A_i(t)$ is the flow of aid from donor country i. In the case of two donor countries, under certain restrictions on parameter values, there exist two symmetric equilibria, one with low aid, and one with high aid. At both equilibria, each country uses a linear Markov-perfect strategy. Interestingly, a lower degree of corruption in the recipient country is associated with a higher low-aid equilibrium, and with a lower high-aid equilibrium. If donor countries are status-conscious, it can be shown that the higher the extent of status-consciousness of the donors, the greater the sum of aids at the low-aid equilibrium, and the smaller the sum of aids at the high-aid equilibrium.

In their second dynamic game model of foreign aid, Kemp and Long (2009) assume that there are two donor countries, and one recipient country. The only state variable is the "level of development" of the recipient country, denoted by $X(t)$. Assume that when X reaches some level $\widehat{X}$, the recipient country's economy can take off and achieve sustained growth without help from abroad. The donor countries want the recipient to achieve the target $\widehat{X}$, and the game ends when this target is reached. Let $A_i(t)$ be the flow of aid from donor country i. Assume that there is an upper bound on aid, such that $0 \leq A_i(t) \leq \overline{A}$.

Starting from any $X < \widehat{X}$, the level of development $X(t)$ evolves according to the following dynamic law

$$\dot{X} = \beta_1(X)A_1 + \beta_2(X)A_2 + \omega(X)A_1 A_2 - \delta(X) \quad \text{for } 0 \leq X < \widehat{X}$$

where $\beta_i(X) > 0$ is the effectiveness of country i's aid. The term $\omega(X) \geq 0$ represents the interactive effect of the two flows of aid. The function $\delta(X)$ represents the depreciation of X. All the functions $\beta_i(.), \omega(.)$ and $\delta(.)$ are differentiable and bounded, for all $X \in [0, \widehat{X}]$. The payoff of donor i is assumed to be

$$J_i = K_i(X(T)) - \int_0^T c_i A_i(t) dt$$

where $c_i > 0$ is the cost per unit of aid, and $K_i(.)$ is the psychological reward at the end of the program. The main results are that the following pair of strategies constitutes an MPNE

$$\phi_i(X) = \frac{\delta(X)}{\beta_i(X)}, \quad i = 1, 2 \tag{4.17}$$

and the value function of donor country i is

$$V_i(X) = K_i(\widehat{X}) - \int_X^{\widehat{X}} \frac{\beta_j(x)c_i}{\beta_1(x)\beta_2(x) + \omega(x)\delta(x)} dx, \quad i = 1, 2 \qquad (4.18)$$

Three points are worth noting:

(i) The strategy (4.17) is, in general, nonlinear in X. For example, consider the following specification of $\widehat{X}$, $\delta(.)$, and $\beta_i(.)$:

$$\widehat{X} = 1$$

$$\delta(X) = 1 - \exp[X - \widehat{X}] \quad \text{for all } X \in [0, \widehat{X}]$$

$$\beta_i(X) = \alpha_i + \delta(X)$$

where $\alpha_i > 0$. Then it can be shown that $\phi_i'(X) < 0$ and $\phi_i''(X) < 0$, that is, as the recipient country's level of development grows, aid from each donor falls at a faster and faster rate. This reflects the fact that, at the end of the horizon, the shadow price of the state variable has a value of zero.

(ii) The equilibrium growth rate of the stock X is

$$\dot{X} = \beta_1(X)A_1^* + \beta_2(X)A_2^* + \omega(X)A_1^*A_2^* - \delta(X)$$

$$= \delta(X) + \omega(X)\frac{[\delta(X)]^2}{\beta_1(X)\beta_2(X)} \qquad (4.19)$$

until $\widehat{X}$ is attained.

(iii) An increase in corruption can be represented as a downward shift in the function β_i. It follows from (4.19) that the higher is the level of corruption, the greater is the flow of aid, and the greater is the growth rate of the stock X (unless $\omega(X) = 0$, so that the growth rate of the stock is independent of the level of corruption).

4.6.2. *Capital flights as a result of rent-seeking by powerful groups*

Tornell and Velasco (1996) modify the fish-war model of Levhari and Mirman (1980) by considering the exploitation of a renewable resource with a linear growth function, and by reinterpreting it as a model of corruption by powerful groups in a developing economy. The model is capable of explaining why capital flows from poor countries to rich countries. The idea is developed further by Tornell and Lane (1999) who use the same model, but offer more intuition. They show that the MPNE leads to slow growth and a "voracity effect" by which a shock, such as a terms-of-trade windfall, perversely generates a more-than-proportional increase in fiscal redistribution and reduces growth. A dilution of the concentration of power leads to faster growth and lower voracity.

Corruption, capital flight, and the voracity effect

The basic model offered by Tornell and Velasco (1996) is as follows. A country has a stock of renewable resource k. There are n powerful groups that exploit this resource. Let $c_i(t)$ be the rate of extraction by group i at time t. The objective function of group i is to maximize its infinite-horizon payoff function

$$\int_0^\infty \left(\frac{\sigma}{\sigma-1}\right) [c_i(t)]^{(\sigma-1)/\sigma} \exp(-\delta t) dt$$

subject to

$$\dot{k}(t) = Ak(t) - c_i(t) - \sum_{j \neq i} c_j(t), \quad A \geq 0$$

where $c_i = 0$ if $k = 0$. Here A is the natural growth rate of the resource, and $\sigma > 0$ is the intertemporal elasticity of substitution. If $\sigma > 1$ then utility is positive and unbounded above. If $0 < \sigma < 1$ then utility is bounded above by zero.

We look for a Markov-perfect equilibrium in linear strategies.[17] Suppose agent i believes that all other agents $j \neq i$ use a linear feedback strategy $c_j = \alpha_j k$. Let

$$\tilde{\beta} \equiv \sum_{j \neq i} \alpha_j$$

Without loss of generality, write $c_i(t) = \alpha_i(t) k(t)$. Then agent i's optimization problem is to find a time path $\alpha_i(t) \geq 0$ that maximizes

$$\int_0^\infty \left(\frac{\sigma}{\sigma-1}\right) k^{(\sigma-1)/\sigma} [\alpha_i(t)]^{(\sigma-1)/\sigma} \exp(-\delta t) dt$$

subject to

$$\dot{k}(t) = [A - \tilde{\beta} - \alpha_i(t)] k(t)$$

---

[17]For results on the class of common property resource games that admit linear Markov strategies, see Long and Shimomura (1998) and Gaudet and Lohoues (2008).

The HJB equation is

$$\delta V_i(k) = \max_{\alpha_i} \left\{ \left(\frac{\sigma}{\sigma - 1} \right) [k\alpha_i]^{(\sigma-1)/\sigma} + V_i'(k)(A - \tilde{\beta} - \alpha_i)k \right\}$$

Let us conjecture that the value function takes the form

$$V_i(k) = \frac{z\sigma}{\sigma - 1} k^{(\sigma-1)/\sigma}$$

where z is to be determined. Maximization of the RHS of the HJB equation gives the first order condition.

$$\alpha_i = z^{-\sigma}$$

Substituting this equation into the HBJ equation, one gets

$$\alpha_i = z^{-\sigma} = \sigma\delta + (1 - \sigma)(A - \tilde{\beta})$$

Thus if $\sigma < 1$ (the bounded utility case), we find that $\tilde{\beta}$ is a strategic substitute for α_i, in the sense that if a player's rivals increase their intensity of exploitation, it will reduce its own intensity (so as to conserve the resource); while if $\sigma > 1$, $\tilde{\beta}$ is a strategic complement of α_i. In the case $\sigma = 1$ (logarithmic utility), we have the dominant strategy $\alpha_i = \delta$ regardless of $\tilde{\beta}$.

In the special case where $n = 1$, we have $\tilde{\beta} = 0$, hence to obtain a positive constant α_i we need to assume that

$$\sigma\delta + (1 - \sigma)A > 0$$

Let us focus on the symmetric equilibrium, that is, all agents choose the same intensity of extraction. The Nash equilibrium is

$$\alpha^N = \frac{\sigma\delta + (1 - \sigma)A}{n - \sigma(n - 1)}$$

where we assume that $n - \sigma(n - 1) > 0$. This means that σ cannot be too large. Clearly, if $\sigma < 1$, an increase in the number of players will reduce α^N. If $\sigma > 1$, an increase in n will increase α^N. The aggregate extraction is $n\alpha^N$, which increases with n.

Tornell and Velasco also develop a more general version of the model, by allowing individuals to hold private wealth (e.g., in a foreign bank account) that yields a constant rate of return $r > 0$. Assume $A > r$. Each agent i can then extract $d_i(t)$ from the common property asset, and deposit it in his private bank account.[18] Consumption $c_i(t)$ is financed by withdrawing from the private account. The balance of this bank account is $f_i(t)$. Thus, each agent faces two differential equations

$$\dot{k}(t) = Ak(t) - d_i(t) - \sum_{j \neq i} d_j(t)$$

$$\dot{f}_i(t) = rf_i(t) + d_i(t) - c_i(t)$$

Assume that there are exogenous upper and lower bounds on $d_i(t)$:

$$\theta_\text{L} k(t) \leq d_i(t) \leq \theta_\text{H} k(t)$$

Tornell and Velasco find that there are three symmetric equilibria. First, there is an interior equilibrium where all players use the extraction strategy

$$d_i(t) = \beta^\text{int} k(t)$$

where $\theta_\text{L} < \beta^\text{int} < \theta_\text{H}$. In the pessimistic equilibrium, everyone extracts at the maximum rate: $d_i(t) = \theta_\text{H} k(t)$. In the optimistic equilibrium, extraction is at the lowest possible rate: $d_i(t) = \theta_\text{L} k(t)$.

To find a symmetric equilibrium, let us conjecture that the value function is of the form

$$V_i(k, f_i) = \frac{z\sigma}{\sigma - 1} (k + f_i)^{(\sigma-1)/\sigma}$$

Solving the HJB equation, we find that

$$\beta^\text{int} = \frac{A - r}{n - 1} > 0$$

provided that $\theta_\text{L} < (A - r)/(n - 1) < \theta_\text{H}$. The equilibrium consumption strategy is

$$c_i(t) = [\delta\sigma + r(1 - \sigma)](k(t) + f_i(t)) \equiv z^{-\sigma}(k(t) + f_i(t))$$

This model shows that capital can flow from poor countries, where the rate of return is high $(A > r)$ to rich countries where the rate of return is lower (r). The reason is that each powerful group knows that while the

---

[18]If $d_i(t) < 0$, this signifies that the individual transfers funds from his private account to the common asset (if this goes on forever, $f_i(t)$ will eventually be negative, which the foreign bank would permit only if it can one day get hold of a fraction of k).

gross rate of return of holding asset in the form of k is A, the net rate of return is only $A - (n-1)\beta^{\text{int}}$, because it faces $(n-1)$ rival powerful groups who can appropriate parts of the common return.

In the pessimistic equilibrium, everyone will extract at the maximum rate, because each powerful group happens to believe that all other powerful groups extract at the maximum rate. In this case, the value function has the following form

$$V_i(k, f_i) = \frac{z\sigma}{\sigma - 1}(qk + f_i)^{(\sigma - 1)/\sigma}$$

where q and z are to be determined. Solving the HJB equation, we get

$$0 < q = \frac{\theta_{\text{H}}}{A - r + n\theta_{\text{H}}} < 1$$

The extraction strategy is

$$d_i(t) = \theta_{\text{H}}k(t)$$

and the consumption strategy is

$$c_i(t) = [\delta\sigma + r(1 - \sigma)](qk(t) + f_i(t)) \equiv z^{-\sigma}(qk(t) + f_i(t))$$

Tornell and Lane (1999) observe that the interior equilibrium of this model display what they call "the voracity effect": an increase in A will lead to an increase in β^{int}. This means that if the poor country under consideration experiences technical progress (or perhaps an improvement in the terms of trade), the powerful groups will extract more, leading to faster depletion of the common asset.

Extensions of the corruption model

In the models studied by Tornell and Velasco (1992) and Tornell and Lane (1999), the extraction from the common property asset stock is costless and the authors claim that "including appropriation or adjustment costs would add nothing to the insights provided by the model" (1999). However Sorger (2005) and Long and Sorger (2006) argue that an explicit consideration of the costs of appropriation is important. A model that takes appropriation costs into account is not only more realistic (after all, money laundering and lobbying involve the use of real resources) but also likely to yield new insights and modify results from Tornell and Velasco (1992) and Tornell and Lane (1999) in non-trivial ways. Long and Sorger (2006) show that both an increase in the appropriation cost and, when appropriation costs vary. across agents, an increase in the degree of heterogeneity of these costs reduce the growth rate of the public capital stock. Thus, they obtain the striking result that high costs of money laundering are detrimental to economic growth.

Another feature of the model of Long and Sorger (2006), which is not present in Tornell and Velasco (1992) and Tornell and Lane (1999), is that the agents derive utility not only from consumption but also from wealth. Wealth is a vehicle for achieving social status, and people do care about social status. Cole *et al.* (1992) argue forcefully that status seeking is a strong motive for economic agents, especially if goods are not allocated through well-functioning markets as it is the case in rent-seeking models. Unlike Tornell and Lane (1999) and Tornell and Velasco (1992), Long and Sorger (2006) do not rely on isoelastic utility functions. The functional form of the utility function can be arbitrary, as long as it possesses concavity and homogeneity of degree one in (k, c). Long and Sorger find that an increase in the degree of heterogeneity of cost leads to slower growth, and under certain conditions, a higher elasticity of substitution between wealth and consumption will lead to a higher intensity of extraction, and thus lower growth.

The above-mentioned models of the tragedy of the commons have a common feature: the assumption that agents care only about their absolute consumption levels (and possibly absolute wealth levels). While useful as a simplifying device, that assumption has recently been criticized because of strong empirical evidence that individuals care about relative consumption (or relative income) as well as absolute consumption: a person's happiness depends on the comparison of his consumption level with that of other members of his peer group. An individual is happier the more his consumption (or income) level exceeds the average consumption (or income) of his reference group. The following question then arises: if agents who exploit a common property renewable resource (either in the literal sense, or in the figurative sense) care about their *relative* performance (consumption, income, or other indices of status), would social welfare and the growth rate of the public asset be more adversely affected compared to the case where they care only about their absolute performance? This question is dealt with in Long and Wang (2009).

Think of a lake that is effectively a common property to a number of municipalities, or provinces. Suppose the leader of each municipality is rewarded according to some relative performance criterion, such as relative employment levels or relative local (municipal) gross domestic product

(GDP) growth rates. Would these leaders have stronger incentive to allow local businesses to pollute the lake? Long and Wang (2009) explore the effect of the concern for relative consumption (or relative income) on the tragedy of the commons, both in the sense of common access natural resources, and in the sense of rent-seeking fiscal appropriations. They take as starting point the model of Tornell and Lane (1996, 1999), where powerful groups grab revenue from a common-access resource, and assume that agents gain utility from both absolute consumption and relative consumption. Long and Wang (2009) also consider the case where agents differ with respect to some characteristics. They introduce two sources of heterogeneity: agents may differ with respect to the degree of status-consciousness as well as with respect to appropriation cost. They show that social welfare decreases if agents become more heterogeneous in terms of status-seeking, but it increases if they become more heterogeneous in terms of appropriation costs.

Chapter 5

DYNAMIC GAMES IN INDUSTRIAL ORGANIZATION

The theory of industrial organization has benefited much from dynamic game-theoretic models of interactions among firms, as well as those between firms on the one hand and far-sighted consumers on the other. In Section 5.1, we review models of dynamic duopoly where the state variables are maybe sticky prices, or sticky quantities, or capital stocks. Section 5.2 considers some extensions of the dynamic oligopoly models. Section 5.3 turns to models of R&D races and technology adoption.

5.1. Dynamic Oligopoly

Economists, in their analysis of oligopoly, have mostly restricted attention to static models. One justification for this lies in the fact that static models are simpler. Another justication is the belief that some static models may be interpreted as a shortcut that basically "captures long-run stable dynamic interactions, with an implicit dynamic game in the background" (Figuières, 2009). This belief, however, turns out to be too optimistic. In many cases, it is true that the steady state of an open-loop Nash equilibrium corresponds to the equilibrium of a static game (especially when some parameters of the dynamic game take on some limiting values). However, MPNE typically yield steady states that are markedly different from those of open-loop Nash equilibria, and one may argue that the feedback strategies capture best the dynamic interactions among firms.

The simplest model of dynamic oligopoly is the so-called "sticky price oligopoly", analyzed in great detail by Fershtman and Kamien (1987). This model has only one state variable, namely the sticky price. However, because of its simplicity, it has been very influential and has spawned a stream of applications and extensions. Reynolds (1987), Driskill and McCafferty (1989), Jun and Vives (2004), and Figuières (2009) have applied differential games to the study of more complicated models of dynamic duopoly, some involving two state variables. They found that dynamic models can yield results about steady-state behavior that are contrary to what one may be led to deduce from the intuition obtained from a static model. For example, using a static model of n Cournot oligopolists, Salant *et al.* (1983) conclude that if members of a subset of Cournot rivals (say m firms) merge into a single firm, such a merger is not profitable unless the ratio m/n is very large (80% or more). This result is no longer true when firms play a dynamic game and use Markov-perfect strategies.

5.1.1. *The sticky price model*

Fershtman and Kamien (1987) consider a duopoly producing a homogeneous good. Let $q_i(t)$ denote the output of firm i at time t, and the total supply is $S(t) = q_1(t) + q_2(t)$. Given the price $p(t)$, the market demand is $D(t) = a - p(t)$. For some reasons, the price is sticky and does not adjust instantaneously to equate demand with supply. It is assumed that price adjustments are sluggish, and the rate of change in price is proportional to excess demand:

$$\dot{p}(t) = \theta[D(t) - S(t)] = \theta[a - p(t) - q_1(t) - q_2(t)] \tag{5.1}$$

Each firm i takes the law of price adjustments, Eq. (5.1), as given, and seeks to maximize the integral of its stream of discounted profit. The firm's production cost is $C_i(q_i) = c_i q_i - \frac{1}{2}q_i^2$.

If firms use open-loop strategies, each firm i chooses a time path of its output $q_i(t)$, taking the time path of its rival's output, $q_j^{\text{OL}}(t)$, as given. Its optimization problem is the following optimal control problem

$$J_i = \max_{q_i} \int_0^\infty e^{-rt}[p(t)q_i(t) - C_i(q_i(t))]dt$$

subject to

$$\dot{p}(t) = \theta[a - p(t) - q_i(t) - q_j^{\text{OL}}(t)], \quad p(0) = p_0 \text{ (given)} \tag{5.2}$$

The state variable of this optimization problem is p, and the control variable is q_i. The dynamic equation (5.2) has an exogenous function of time $q_j^{\text{OL}}(t)$ on the RHS. This is, therefore, a non-autonomous optimal control problem. An open-loop Nash equilibrium is a pair of time paths of output $\{\tilde{q}_1^{\text{OL}}(.), \tilde{q}_2^{\text{OL}}(.)\}$ that are best replies to each other. That is,

$$J_1\{\tilde{q}_1^{\text{OL}}(.), \tilde{q}_2^{\text{OL}}(.)\} \geq J_1\{q_1^{\text{OL}}(.), \tilde{q}_2^{\text{OL}}(.)\} \quad \text{for all } q_1^{\text{OL}}(.) \neq \tilde{q}_1^{\text{OL}}(.)$$

$$J_2\{\tilde{q}_1^{\text{OL}}(.), \tilde{q}_2^{\text{OL}}(.)\} \geq J_2\{\tilde{q}_1^{\text{OL}}(.), q_2^{\text{OL}}(.)\} \quad \text{for all } q_2^{\text{OL}}(.) \neq \tilde{q}_2^{\text{OL}}(.)$$

If firms use feedback strategies, each firm i formulates a decision rule that determines its output at any time t as a function of the observed level of the state variable at that time:

$$q_i(t) = \phi_i^{\text{FB}}(p(t))$$

Given the rival's decision rule $\phi_j^{\text{FB}}(p)$, firm i's optimization problem is an autonomous optimal control problem:

$$J_i = \max_{q_i} \int_0^\infty e^{-rt}[p(t)q_i(t) - C_i(q_i(t))]\, dt$$

subject to

$$\dot{p}(t) = \theta[a - p(t) - q_i(t) - \phi_j^{\text{FB}}(p)], \quad p(0) = p_0 \text{ (given)}$$

A feedback Nash equilibrium is a pair of decision rules for output $\{\widetilde{\phi}_1^{\text{FB}}(.), \widetilde{\phi}_2^{\text{FB}}(.)\}$ that are best replies to each other. That is,

$$J_1\{\widetilde{\phi}_1^{\text{FB}}(.), \widetilde{\phi}_2^{\text{FB}}(.)\} \geq J_1\{\phi_1^{\text{FB}}(.), \widetilde{\phi}_2^{\text{FB}}(.)\} \quad \text{for all } \phi_1^{\text{FB}}(.) \neq \widetilde{\phi}_1^{\text{FB}}(.)$$

$$J_2\{\widetilde{\phi}_1^{\text{FB}}(.), \widetilde{\phi}_2^{\text{FB}}(.)\} \geq J_2\{\widetilde{\phi}_1^{\text{FB}}(.), \phi_2^{\text{FB}}(.)\} \quad \text{for all } \phi_2^{\text{FB}}(.) \neq \widetilde{\phi}_2^{\text{FB}}(.)$$

Note that the argument of the functions ϕ_1^{FB} and ϕ_2^{FB} is p, not t.

For simplicity, assume that the firms are symmetric, that is, $c_1 = c_2 = c$. Then it can be shown that an open-loop equilibrium exists, and that along the equilibrium path, the state variable p converges to the steady-state value

$$p_\infty^{\text{OL}} = \frac{a(2\theta + r) + 2c(r + \theta)}{3r + 4\theta}$$

Notice that if $r \to 0$ or $\theta \to \infty$, the steady-state value p_∞^{OL} will be the same as the static Cournot equilibrium price,

$$p^{\text{C}} = \frac{a + c}{2}$$

Interestingly, in the opposite case where $r \to \infty$ or $s \to 0$, the steady-state value p_∞^{OL} will be the same as the perfect competition price, where each firm's output is large enough to equate its marginal cost to the price.

Let us turn to the case where firms use feedback strategies. Since the problem is linear quadratic, we focus on the linear feedback strategies of the form

$$q_i = \phi_i^{\text{FB}}(p) = \alpha_i p + \beta_i$$

where α_i and β_i are to be determined.[1] For simplicity, assume $c_1 = c_2 = c$. Then it can be shown that in equilibrium the firms use the same linear feedback strategy, with

$$\alpha_1 = \alpha_2 = \alpha = (1 - \theta K) > 0$$

$$\beta_1 = \beta_2 = \beta = \theta E - c < 0$$

where

$$K = \frac{r + 6\theta - \sqrt{(r + 6\theta)^2 - 12\theta^2}}{6\theta^2} > 0$$

and

$$E = \frac{-a\theta K + c - 2\theta cK}{r - 3\theta^2 K + 3\theta}$$

(Here, the positive root K is chosen to ensure the convergence of p to a steady state.)

Notice that $\alpha > 0$, which indicates that the feedback decision rules tell firms to produce more if they observe a higher price. This suggests that in the feedback equilibrium, firms are more competitive than in the open-loop equilibrium. Let us verify this. Computing the steady-state price under the feedback equilibrium, we get

$$p_\infty^{\mathrm{FB}} = \frac{a + 2c - 2\theta E}{3 - 2\theta K}$$

In the limiting case, where $r \to \infty$ or $\theta \to \infty$, we can see that $p_\infty^{\mathrm{FB}} < (a+c)/2$ (which is the static Cournot price, which is identical to the steady-state open-loop Nash equilibrium price under the stated conditions). The reason is that if each firm expects its rival to produce more when the price is higher then it has little incentive to restrict output. The feedback equilibrium therefore corresponds to the static model with negative conjecture. This intuition has been stressed by Dockner (1992) and Dryskill and McCafferty (1989).

The sticky price model has been applied to the analysis of international trade under duopolistic competition (Dockner and Haug, 1990, 1991) and mergers (Dockner and Gaunerdorfer (2001); Benchekroun (2003)). Driskill and McCafferty (1996) use the sticky-price model to reexamine the third

---

[1]Tsutsui and Mino (1990) show that in addition to the equilibrium linear strategies mentioned below, there is a continuum of feedback Nash equilibria where the firms use non-linear strategies.

market model of static international duopoly. Fujiwara (2010) extends the Dockner–Haug model to the case where the number of firms in H and F are not the same. An important and influential extension of the Fershtman–Kamien analysis is provided by Tsutsui and Mino (1990), who show the existence of a continuum of feedback equilibria where the duopolists use non-linear strategies. A subset of these strategies generates higher steady-state profits for the firms relative to the steady-state profits found by Fershtman and Kamien.

5.1.2. *Investment in capacity by Cournot oligopolists*

Reynolds (1987), and Driskill and McCafferty (1989) consider a dynamic duopoly where firms accumulate their capital stocks.[2] The following formulation presents their basic models, in the light of later syntheses offered by Jun and Vives (2004), and Figuières (2009).

We consider a simple linear quadratic game with two firms, each building its own capital stock. Let $K_i(t)$ $(i = 1, 2)$ denote the capital stock of firm i at time t. Let $u_i(t)$ denote the control variable of firm i; we may think of u_i as firm i's investment at time t. Time is a continuous variable. The following differential equation describes how K_i evolves over time:

$$\dot{K}_i(t) = u_i(t) - \delta K_i(t), \quad K_i(0) = K_{i0}$$

where $\delta \geq 0$ is the rate of depreciation. Assume that one unit of capital produces one unit of output, and assume that capital is fully employed. Cournot competition under linear demand yields *instantaneous gross payoffs* that are quadratic functions of the capital stocks. We denote the *instantaneous gross payoffs* for firm i $(i = 1, 2)$ by

$$R_i(K_i, K_j) = a + b_o K_i + b_r K_j + \frac{m_o}{2} K_i^2 + \frac{m_r}{2} K_j^2 + m_x K_i K_j$$

where the subscript o in b_o and m_o indicates the parameters associated with the firm's "own capital stock", and the subscript r in b_r and m_r indicates the parameters associated with the "rival's capital stock". The subscript x in m_x indicates the "cross effect".

Recall the static analysis of oligopoly by Bulow *et al.* (1985). They introduced the concepts of "strategic substitutes" and "strategic

---

[2]An earlier paper by Spence (1979) deals with a continuous time accumulation game by oligopolists. He considers only open-loop strategies.

complements". In their relatively simple context, a firm's choice variable is a strategic substitute to its rival's choice variable if its reaction function is downward sloping. In a dynamic oligopoly game, the ideas of "strategic substitutes" and "strategic complements" are not so straightforward, because each firm has a control variable (e.g., its rate of investment) and a state variable (e.g., its capital stock). We will need three different definitions of substitutability (and complementarity), which we introduce sequentially as we proceed with our analysis, and we discuss the relationship among these definitions, and their implications.

Definition 1 The model is said to display *"instantaneous gross payoff substitutability"* (respectively, complementarity) if and only if the cross effect is negative, that is, the marginal contribution of a player's state variable (here, capital) to its instantaneous gross payoff is a decreasing (respectively, increasing) function of the rival's state variable.

Since $R_i(K_i, K_j)$ is quadratic, *instantaneous gross payoff substitutability* holds if and only if $m_x < 0$.

Assumption A1

$$a \geq 0, \quad b_o > 0, \quad m_o < 0 \quad \text{and} \quad m_r \leq 0$$

This assumption ensures that R_i is concave in each of the capital stock, and that each firm's payoff is increasing in its own capital stock when the latter is small.

When a firm makes an investment u_i, it incurs a quadratic cost, represented by the following equation:

$$C(u_i) = cu_i + \frac{e}{2}u_i^2 \quad \text{where } c \geq 0 \text{ and } e > 0 \tag{5.3}$$

We call $R_i(K_i, K_j) - C(u_i)$ is the *instantaneous net payoff of firm i*.

Firm i's lifetime payoff is the integral of the flow of *instantaneous net payoff*, discounted at the rate $\rho > 0$:

$$J_i = \int_0^\infty \exp(-\rho t)[R_i(K_i, K_j) - C(u_i)]\, dt$$

A firm's open-loop strategy is a function of time only. It tells us the time path of the firm's control variable u_i:

$$u_i(t) = \phi_i(t)$$

The function $\phi_i(t)$ in general depends on the initial capital stocks and may be written as $\phi_i(t, K_{i0}, K_{j0})$. It is independent of the observed values $K_i(t)$

and $K_j(t)$ for $t > 0$. In this sense, open-loop strategies are "insensitive" to deviations of the state variables from their projected time paths. We assume that the functions $\phi_i(.)$ $(i = 1, 2)$ are piece-wise continuous.

In contrast, a feedback strategy is dependent on the state variables. A feedback strategy is said to be stationary if it does not depend on time. We restrict attention to stationary feedback strategies:

$$u_i(t) = \phi_i^{\text{FB}}(K_i(t), K_j(t))$$

Here, the function ϕ^{FB} depends only on the state variables K_i and K_j, and not on time. An MPNE is a pair of feedback strategies such that each firm's strategy is a best response to the other firm's strategy, for any state vector (K_1, K_2). Hence MPNE possess a desirable property that corresponds to the "subgame perfect" requirement. We restrict attention to feedback strategies that are continuous and almost everywhere differentiable in the state variables.

Remark 1 (A different interpretation of the model.) If we set $a = \delta = b_r = 0$ and $c = 0$, we can reinterpret the model as one of **Cournot duopoly with costly output adjustments**. Here K_i is the output (not capital) of firm i, and u_i is the rate of change in its output. The inverse demand function is $P = b_o + m_o(K_1 + K_2)$, where $m_o < 0$, and $m_x = m_o$.

Co-operative accumulation of capital stocks

Consider the case where, despite their rivalry in the product market, both firms coordinate their capacity-building strategies so as to maximize the sum of their payoffs, $J_1 + J_2$. The solution of this joint maximization problem can be found using the standard technique of optimal control theory (e.g., Leonard and Long, 1992).

Assuming that the parameters satisfy the following inequalities

$$b_o + b_r > (\rho + \delta)c$$

$$(\rho + \delta)\delta e - (m_o + m_x) - (m_r + m_x) > 0$$

one can show that the co-operative equilibrium leads to a positive steady-state $(\overline{K}_1^*, \overline{K}_2^*)$ where

$$\overline{K}_1^* = \overline{K}_2^* = \frac{b_o + b_r - (\rho + \delta)c}{(\rho + \delta)\delta e - (m_o + m_x) - (m_r + m_x)} > 0$$

With some additional restrictions, one can show that for any initial conditions (K_{10}, K_{20}) there exists a unique corresponding investment path (u_1, u_2) such that (K_1, K_2) converges to the steady state.

Open-loop Nash equilibrium of the capacity-building game

Now assume that the players do not co-operate, and that they use open-loop strategies. To solve for an OLNE, one sets up two optimal control problems, one for each firm, and imposes the restriction that each firm's optimal time path is taken by the rival firm as an exogenous path. It can be shown that there exists a unique OLNE that leads to a unique steady state given by

$$\overline{K}_1^{\text{OL}} = \overline{K}_2^{\text{OL}} = \frac{b_{\text{o}} - (\rho + \delta)c}{(\rho + \delta)\delta e - (m_{\text{o}} + m_x)}$$

If we assume that $(m_{\text{r}} + m_x) < 0$ and $b_{\text{r}} < 0$ then the OLNE steady-state stock is greater than the co-operative solution (i.e., there is *overinvestment* in capacity). Fershtman and Muller (1984) deal with this open-loop case, and extend it to the case of OLNE with asymmetric firms.

Remark 2 Consider the special case where $\delta = 0$ and $c = 0$. As mentioned earlier, we can interpret K_i as the output of firm i and u_i as the rate of change in output. This interpretation gives us a model of Cournot duopoly with costly output adjustments. In this case, the OLNE steady state becomes

$$\overline{K}_1^{\text{OL}} = \overline{K}_2^{\text{OL}} = \frac{b_{\text{o}}}{-(m_{\text{o}} + m_x)}$$

which is identical to the static Cournot equilibrium. It follows that a justification of the study of static models is that they yield the steady state of an OLNE. However, as we have pointed out, OLNE is generally not "credible". Generically, one can show that the steady state of an MPNE differs from that of an OLNE, and this indicates that static models can be misleading.

Markov-perfect Nash equilibrium of the capacity-building game

To solve for an MPNE of the game under consideration, a standard approach is to find a pair of feedback strategies $\phi_1^{\text{FB}}(K_1, K_2)$ and

$\phi_2^{\mathrm{FB}}(K_2, K_1)$ and value functions $V_1(K_1, K_2)$ and $V_2(K_2, K_1)$ such that:

(i) $\phi_1^{\mathrm{FB}}(K_1, K_2)$ is the solution of the HJB equation:

$$\rho V_1(K_1, K_2) = \max_{u_1} \left[R_1 - C(u_1) + \frac{\partial V_1}{\partial K_1}(u_1 - \delta K_1) \right.$$
$$\left. + \frac{\partial V_1}{\partial K_2}(\phi_2^{\mathrm{FB}} - \delta K_2) \right]$$

(ii) $\phi_2^{\mathrm{FB}}(K_2, K_1)$ is the solution of

$$\rho V_2(K_2, K_1) = \max_{u_2} \left[R_2 - C(u_2) + \frac{\partial V_2}{\partial K_1}(\phi_1^{\mathrm{FB}} - \delta K_1) \right.$$
$$\left. + \frac{\partial V_1}{\partial K_2}(u_2 - \delta K_2) \right]$$

(iii) the resulting time path $(K_1(t), K_2(t))$ converges to a steady state $(\overline{K}_1^{\mathrm{FB}}, \overline{K}_2^{\mathrm{FB}})$.

We now introduce our second concept of substitutability (and complementarity).

Definition 2 The value function V_i is said to display *"value function substitutability"* (respectively, complementarity) if and only if $\partial^2 V_i / \partial K_i \partial K_j < 0$.

Since our problem is linear quadratic and the payoff functions are symmetric, it is natural to look for linear feedback strategies of the form

$$u_1(t) = \phi_1^{\mathrm{FB}}(K_1(t), K_2(t)) = A + B_{\mathrm{o}} K_1(t) + B_{\mathrm{r}} K_2(t) \qquad (5.4)$$

$$u_2(t) = \phi_2^{\mathrm{FB}}(K_2(t), K_1(t)) = A + B_{\mathrm{o}} K_2(t) + B_{\mathrm{r}} K_1(t) \qquad (5.5)$$

and quadratic value functions of the form

$$V_1(K_1, K_2) = \theta + v_{\mathrm{o}} K_1 + v_{\mathrm{r}} K_2 + \frac{w_{\mathrm{o}}}{2} K_1^2 + \frac{w_{\mathrm{r}}}{2} K_2^2 + w_x K_1 K_2 \qquad (5.6)$$

$$V_2(K_2, K_1) = \theta + v_{\mathrm{o}} K_2 + v_{\mathrm{r}} K_1 + \frac{w_{\mathrm{o}}}{2} K_2^2 + \frac{w_{\mathrm{r}}}{2} K_1^2 + w_x K_1 K_2 \qquad (5.7)$$

The value of the vector of unknowns $(\theta, v_{\mathrm{o}}, v_{\mathrm{r}}, w_{\mathrm{o}}, w_{\mathrm{r}}, w_x)$ can be determined from the parameters of the instantaneous gross payoff function $R_i(K_i, K_j)$, the control cost function $C(u_i)$, and the discount rate ρ.

Having found the value functions $V_1(K_1, K_2)$ and $V_2(K_2, K_1)$ as in Eqs. (5.6) and (5.7), one can show that the feedback strategies (5.4) and (5.5) satisfy

$$A = \frac{v_o - c}{e}, \quad B_o = \frac{w_o}{e}, \quad B_r = \frac{w_x}{e}$$

Furthermore, the paths of the state variables converge to the steady states defined by

$$\delta \overline{K}^{\mathrm{FB}} = A + (B_o + B_r)\overline{K}^{\mathrm{FB}}$$

The feedback strategies (5.4) and (5.5) show that a firm's optimal rate of investment, u_i, depends on both its current stock level K_i and on the rival's stock level K_j. This leads us to the third definition of the concepts of substitutability and complementarity.

Definition 3 The feedback strategy $u_i = \phi_i^{\mathrm{FB}}(K_i, K_j)$ is said to display *Markov control-state substitutablity* (respectively, *Markov control-state complementarity*) if and only if $\partial \phi_i^{\mathrm{FB}}/\partial K_j < 0$ (respectively, > 0).

Without calculating B_r, can we determine its sign from the parameters of the model? The answer is provided by the following result.

Proposition 1 *Assume that the adjustment cost function is of the form (5.3). Then*

(i) Markov control-state substitutability, $B_r < 0$, (respectively, Markov control-state complementarity, $B_r > 0$) holds if and only if the game displays *"instantaneous gross payoff substitutability"* (respectively, complementarity), that is, iff the coefficient m_x is negative (respectively, positive).

(ii) Value-function substitutability (respectively, complementarity) is equivalent to Markov control-state substitutability (respectively, Markov control-state complementarity).

Remark 3 It will be seen in the next subsection that the equivalence stated in Proposition 1 does not hold if the adjustment cost to a player depends on the control variable of both players.

Proposition 2 *Under Cournot rivalry with costly capacity adjustments, there is a unique symmetric MPNE in linear strategies that leads to a steady state. The steady-state stock is more competitive than the static Cournot equilibrium (i.e., the steady-state stock is **bigger** than the static Cournot output). It is greater than the OLNE steady-state stock which in turn is greater than the fully collusive stock.*

Remark 4 With the interpretation of Cournot competition with costly output adjustments, $\overline{K}^{\text{FB}}$ is the MPNE steady-state output of each firm, and this is greater than the static Cournot equilibrium level. In other words, the overexpansion of capacity (relative to the co-operative outcome) is more pronounced than in the static model. The reason is as follows. Firm i knows that firm j will "invest" less (have a lower u_j) if K_i is higher. So, firm i has an added incentive to build up K_i. This is the dynamic Markov strategic effect, which operates in addition to the usual static "strategic substitute" consideration discussed by Bulow *et al.* (1985).

Proposition 2 has an interesting implication for mergers. Salant *et al.* (1983) show that under linear demand and static Cournot competition, a merger is profitable only if it involves more than 80% of the firms. Does this result carry over to the dynamic oligopoly? The answer is "no". Since MPNE is more competitive than static Cournot rivalry, a merger (in a dynamic model of capacity building) can be profitable even if a smaller fraction of the industry is involved.

5.1.3. *Bertrand rivalry with costly price adjustments*

We now turn to a model of Bertrand rivalry with costly price adjustments. As pointed out by Jun and Vives (2004), it turns out that this can be done by a simple reinterpretation of variables.

Consider the following reinterpretation of the variables. Let us assume that each firm produces a different good, and the two goods are imperfect substitutes. Assume that the demand for good i is

$$D_i = \alpha - \beta P_i + \gamma P_j, \quad \alpha > 0 \tag{5.8}$$

where $\beta > \gamma > 0$. Let q_i denote the output of firm i. Assume that each firm satisfies its demand at each instant, so that $q_i = D_i = \alpha - \beta P_i + \gamma P_j$. The gross profit of firm i is

$$\pi_i = P_i D_i = P_i(\alpha - \beta P_i + \gamma P_j) = \alpha P_i - \beta P_i^2 + \gamma P_i P_j$$

Assume that production costs are zero, and prices are costly to adjust. Let u_i be the rate of price adjustment, that is, $\dot{P}_i = u_i$. Assume that the cost of price adjustment is

$$C(u_i) = cu_i + \frac{e}{2}u_i^2 \quad \text{where } c = 0 \text{ and } e > 0$$

Then we can apply the capital accumulation model in the preceding section to the present model, by replacing K_i by P_i, and set $\delta = 0$, $a = c = 0$,

$b_{\rm o} = \alpha$, $b_{\rm r} = 0$, $m_{\rm o} = -2\beta < 0$, $m_x = \gamma > 0$. Applying Proposition 1 (with suitable reinterpretation of variables), we conclude that this model exhibits Markov control-state complementarity: the higher the rival's price P_j, the greater the firm's own rate of price adjustment, u_i.

Proposition 3 *Under Bertrand rivalry with costly price adjustments, there is a unique symmetric MPNE in linear strategies that leads to a steady state. The steady-state price is less competitive (more collusive) than the static Bertrand equilibrium. That is, the steady-state price is higher than the static Bertrand price. It lies strictly between the OLNE steady-state price and the fully collusive price. (See Jun and Vives, 2004.)*[3]

5.1.4. *Bertrand rivalry with costly output adjustments*

Let us turn to a mixed model (by Jun and Vives, 2004). We continue to assume that prices are the state variables, such that price adjustments are control variables, but specify that a firm's cost of adjustment is not a quadratic function of its price adjustments, rather it is the firm's output adjustment that is costly. More precisely, let us retain the demand function as in Eq. (5.8), and assume that the firm's output is equal to the demand:

$$q_i = D_i = \alpha - \beta P_i + \gamma P_j$$

When both firms adjust their prices, the resulting output changes are

$$\dot{q}_i = -\beta \dot{P}_i + \gamma \dot{P}_j$$

Firm i's cost of adjustment is assumed to be

$$C_i = \frac{\lambda}{2}(\dot{q}_i)^2 = \frac{\lambda}{2}(-\beta \dot{P}_i + \gamma \dot{P}_j)^2$$

This model is more complicated than the previous model, because the instantaneous payoff for each firm i is dependent on the control variable of the other firm, that is, on $\dot{P}_j$, which in turn is a function (via the feedback strategies) of both state variables. Since in this case, firm i's adjustment costs depend on the control variables of both firms, it is no longer true that Markov control-state substitutability implies value function substitutability.

---

[3]Note that for finite-horizon linear-quadratic games, if we insist that strategies be analytic functions of the state variables then there is only one MPNE: the linear strategy pair (Jun and Vives, footnote 13).

Jun and Vives (2004) show that there exists a unique OLNE that leads to a stable steady state. They also find that there exists a unique stable linear MPNE. In this mixed model of Bertrand rivalry, Jun and Vives find that the MPNE is *more competitive* than the OLNE. This is because in the presence of costly output adjustment, the *instantaneous payoff complementarity* of the static Bertrand game is turned into the *Markov control-state substitutablity* in the dynamic game.

5.1.5. *Going beyond the linear quadratic model*

Jun and Vives (2004) extend the results on Markov control-state substitutes and Markov control-state complements to models that are not linear quadratic. They consider a duopoly game with two state variables, say y_1 and y_2, where each firm controls the rate of change of one state variable. Suppose $\dot{y}_i = u_i$, and u_i is the control variable of firm i. Assume each firm's instantaneous payoff function is the sum of two functions: (a) the first function, R_i, depends only on the two state variables, and (b) the second function is an adjustment cost function, which depends only on the *rate of change* of one (or both) state variables. Define z_i as some aggregator function of the two state variables: $z_i = z_i(y_1, y_2)$. Then, $\dot{z}_i = (\partial z_i/\partial y_1)\dot{y}_1 + (\partial z_i/\partial y_2)\dot{y}_2$. The second function is denoted by $F(\dot{z}_i)$. Assume R_i is concave in y_i, and F is convex in $\dot{z}_i$.

If $\dot{z}_i = u_i \equiv \dot{y}_i$, we call this "the pure adjustment case". If $\dot{z}_i$ depends on both u_i and u_j, we call this the "mixed case". In the "pure case", it remains true that, even for non-linear quadratic models, "control-state Markov substitutability" holds if and only if "value function substitutablity" holds. In the mixed case, in general nothing can be said about the relationship between value-function substitutability and control-state substitutability.

Other models of oligopoly that do not rely on the linear quadratic functional forms include Koulovatianos and Mirman (2007) and Dockner and Mosburger (2007). The latter paper makes use of numerical solution, except for a special case.[4]

5.2. Extensions of Dynamic Oligopoly Games

In this section, we discuss two extensions of dynamic oligopoly games. The first extension deals with markets where consumers are farsighted. The second extension includes natural resource dynamics in oligopolistic rivalry.

---

[4]Judd (1998) is a useful work on numerical methods in economics.

5.2.1. *Dynamic oligopoly with far sighted consumers*

In the models of dynamic oligopoly considered in the preceding section it was assumed that consumers behave as if their choices today are independent of future prices. This simplifying assumption is useful in some contexts, but can be very misleading in some other contexts. It was Coase (1972) who pointed out that if consumers purchase durable goods, they must take into account future resale values and/or future implicit rental rates of the durable goods. Coase argued that when consumers of durable goods are farsighted, the correct equilibrium concept requires that no consumers can be fooled by the firms (e.g., any changes in capital values of durable goods must be fully anticipated). Thus, Coase was in fact imposing Markov-perfect behavior on both firms and consumers. The issue was taken up by Stokey (1981), Kahn (1986), and Karp (1996), among others, in the context of supply of a durable good by a monopolist. These authors, however, only dealt with the monopoly case. Furthermore, it becomes clear that the issue of perfect equilibrium when consumers are farsighted is not restricted to the purchase of durable goods. Driskill and McCafferty (2001) explore a model of oligopoly provision of a homogeneous addictive good, given that consumers know they will get addicted if they purchase the good. Laussel *et al.* (2004) consider the purchase of an annual membership of a club when consumers rationally expect the future congestion rates at each of the two clubs they are choosing from.

Oligopolistic provision of an addictive good

Suppose that the consumption of a good (say cigarettes) can cause addiction: the more a person consumes it, the greater is his future demand at any given price. To capture this fact, Driskill and McCafferty (2001) define a person's *"consumption capital"* of the addictive good to be an exponentially weighted average of his past consumption:

$$z(t) \equiv \int_0^t c(\tau) \exp[s(t - \tau)]d\tau, \quad s > 0$$

where s is a positive parameter. Then, the rate of change in his consumption capital is

$$\dot{z} = c - sz$$

Assume that his instantaneous utility at time t depends on $z(t)$, $c(t)$, and $x(t)$ (where the latter is his consumption of a non-addictive good, which serves as the numeraire good):

$$U = \alpha_0 c - \frac{\alpha}{2}c^2 + \beta_0 z - \frac{\beta}{2}z^2 + \delta cz + x$$

where $\delta > 0$.

Consider a world consisting of a continuum of identical consumers that maximize their lifetime (infinite horizon) utility

$$\max \int_0^\infty \exp[-\rho t] \left[\alpha_0 c - \frac{\alpha}{2}c^2 + \beta_0 z - \frac{\beta}{2}z^2 + \delta cz + x \right] dt$$

subject to the budget constraint

$$x(t) + p(t)c(t) = y(t)$$

where $y(t)$ is the consumer's exogeneous income.

Consumers expect that the market equilibrium price at any time t depends only on the state variable $Z(t)$ (the aggregate consumption capital). This dependence is expressed by an endogenously determined price function $P(.)$:

$$p(t) = P(Z(t))$$

The function $P(.)$ depends on all the parameters of the model.

Firms, on the other hand, by inspecting the representative consumer's first-order condition, believe that the price function is of the form

$$p = p(t) = \chi_0 - \alpha C(t) + \chi Z(t) \tag{5.9}$$

where χ_0 and χ are endogenously determined. This linearity is basically derived from the fact that each consumer's shadow price of his addictive capital stock z is linear in z. In a symmetric equilibrium, it holds that $z(t) = Z(t)$. However, while each consumer takes the time path of the aggregate stock $Z(t)$ as given, he thinks that he can influence his own stock $z(t)$ independently of $Z(t)$.

There are a finite number of firms that produce the addictive good. Each firm i knows that in equilibrium $C = Q \equiv \sum q_j \equiv Q_{-i} + q_i$ where q_i is firm i's output. It also knows that

$$Q_{-i} = \sum_{i \neq j} \widehat{q}_j(Z)$$

where $\widehat{q}_j(Z)$ is firm j's production strategy. Then firm i perceives that it can influence the price $p(t)$ by controlling its supply $q_i(t)$

$$p(t) = p(q_i(t) + Q_{-i}(Z(t)), Z(t)) = \chi_0 - \alpha q_i(t) - \alpha Q_{-i}(Z(t)) + \chi Z(t)$$

The firm's revenue function is

$$R_i = p(q_i + Q_{-i}(Z), Z)q_i = [\chi_0 - \alpha q_i(t) - \alpha Q_{-i}(Z) + \chi Z(t)]q_i$$

Let $G(q_i)$ be the total cost of producing q_i. Firm i maximizes the integral of its discounted profit flows:

$$\int_0^\infty \exp[-\rho t][R_i - G(q_i)]dt$$

subject to

$$\dot{Z} = q_i + Q_{-i}(Z) - sZ$$

The Hamiltonian for firm i is

$$H_i = [\chi_0 - \alpha q_i(t) - \alpha Q_{-i}(Z) + \chi Z(t)]q_i - G(q_i) + \lambda_i[q_i + Q_{-i}(Z) - sZ]$$

In equilibrium, consumers' expectations are correct, that is,

$$P(Z) = \chi_0 - \alpha n\widehat{q}_i(Z) + \chi Z$$

It can be verified that there exists a Markov-perfect equilibrium in which firms use a linear feedback strategy of the form

$$q_i(t) = A + hZ(t)$$

and this is consistent with consumers' expectations.

The main results of the analysis of the model are as follows.

(i) In the steady state, marginal cost of production may exceed price. This is because each firm equates its marginal cost with its "full marginal revenue", which includes the effect of its current output on the stock of addictive capital (the addictive term in consumer demand).

(ii) The steady-state profit of a firm is not monotone decreasing in the number of firms. This is because any increase in the number of firms has two opposite effects: (a) more competition (which reduces profits of individual firms), and (b) more free ride (under-investment in the consumption capital).

Dynamic duopoly with anticipated future congestion

Laussel *et al.* (2004) consider a dynamic model of price competition between two profit-maximizing clubs. They study the evolution of access prices and market shares. They postulate two congestible networks that provide horizontally differentiated services, such as two web sites. Once customers have selected a firm, they cannot instantaneously change their mind. Consumers take into account not only the prices and *"distance"* to each firm but also the anticipated future congestion rates, when they make their decision to belong to one or the other "profit-maximizing club".

Two firms are located at the two extremities of a linear city, represented by the unit interval $[0, 1]$. Each firm provides access to its facility of a given size. (Sizes are different between firms.) Consumers are uniformly distributed along the line segment and each has to pay initially a subscription price and to bear at each point of time a quadratic travel cost. A consumer located at $x \in [0, 1]$ bears a "travel cost" of τx^2 per unit of time if he chooses firm 1, and $\tau(1 - x)^2$ if he chooses firm 2. Travel cost may also be interpreted as a utility loss from the divergence between the type of the facility of his club and his "ideal" (most preferred) facility type.

A consumer's instantaneous utility for using one facility is a decreasing function of the level of congestion of that facility, measured by the ratio of the number of customers to the size of the facility. Each consumer faces a constant probability of exiting the market. Further, at each point of time, a flow of new consumers enters the market. Under the assumption that the entry rate equals the exit rate, the population is constant. The firms determine their subscription prices at each point of time. Each firm takes into account the fact that attracting new consumers today by lowering prices will result in larger future congestion rates and hence less new consumers in the future.

Laussel *et al.* (2004) model duopolistic competition as a differential game where the firms' strategies *and the customers' expectations* depend only on the "payoff-relevant state", that is, the market shares. They analyze the MPNE of the game where firms' pricing strategies and consumers' expectations about future congestion are functions of the market shares.

Their main results are as follows. Concerning equilibrium pricing strategies, in the case where the facility sizes are equal, if a firm finds itself with a larger market share than its rival, it will charge a lower price. This is because of a higher congestion rate at this firm which the customers rationally expect to prevail for a while. Concerning steady-state prices, if the sizes of the facilities are equal, steady-state prices are *higher* than what they would be if there were no congestion effects. This is because the presence of future negative consumption externalities softens duopolistic competition in prices. In fact, there is little incentive to cut prices to attract more new consumers when eveyone knows that this will result in higher future congestion rates which will render the firm less attractive for future new consumers. It is rather clear that positive consumption externalities would have the opposite effect. A somewhat surprising implication of this result is that the firms have an incentive to collude on equal reduction of their capacities, because this would result in higher congestion costs, and consequently higher steady-state prices and profits.

5.2.2. *Oligopolistic extraction of natural resources*

Early models of oligopoly in exhaustible resources use the open-loop equilibrium concept.[5] Under constant elasticity of demand and zero marginal cost, a feedback equilibrium can be found (e.g. Benchekroun and Long, 2006).[6] Salo and Tavohnen (2001) characterize the feedback equilibrium in a duopoly in non-renewable resource. They assume that costs rise as the stocks dwindle, and eventually marginal cost becomes so high that firms abandon the deposits without exhausting them. For renewable resources, using a "tent-shaped" growth function and a linear demand function, Benchekroun (2003, 2008) was able to characterize the feedback equilibrium for a duopoly. An alternative assumption about demand is found in Jørgensen and Yeung (1996).

There is a literature that deals with cartels versus fringes in the non-renewable resource market. Salant (1972) characterizes an OLNE between a resource cartel and a fringe. Newbery (1981) discusses the issue of dynamic inconsistency in the a cartel-fringe model of resource extraction. Groot *et al.* (2003) define the concept of strong consistency and apply it to the cartel-fringe model. For more references and discussion of the cartel-fringe model, see Benchekroun and Withagen (2008).

5.3. R&D Races and Technology Adoption

This section deals with two important areas where dynamic games have made significant contributions: R&D races and technology adoption. While

---

[5]See Lewis and Schmalensee (1980), Loury (1986), and Gaudet and Long (1994).
[6]Long *et al.* (1999) find that for a class of differential games, some meaningful comparison of OLNE and MPNE can be obtained easily.

it is possible to model rivalry in R&D expenditures and technology adoption in a static context, these topics are best analyzed in a dynamic framework. Within such a framework, one can study the roles of uncertainty resolution, diffusion of technology, pre-emption, and so on.

5.3.1. *R&D races*

Consider a number of firms that enter a R&D race, with the hope of being the first one to discover a new product (say, a cancer cure). The discoverer will be awarded a prize (say, a patent, which yields monopoly profit). Firms invest in R&D, but the outcomes are uncertain: the time at which the discovery is made is stochastic, and the identity of the winner is also stochastic.

Lee and Wide (1980) assume that (i) firms choose a *constant* R&D effort, and (ii) the (undiscounted) value of the prize is the same regardless of when the discovery is made (hence the *present value of the prize is a decreasing function of the discovery date*). Lee and Wide show how the equilibrium constant R&D effort is dependent on the number of firms and on the hazard rate.

Reinganum (1982a) departs from these assumptions. She does not postulate that R&D efforts are constant. On the other hand, she imposes a new assumption on the prize: that present value of the prize is constant, in other words, the undiscounted prize $\Omega(t)$ is assumed to grow at a constant rate g which happens to be equal to the rate of interest r. This assumption serves to "eliminate" the effect of growth or decline in the prize on the R&D efforts, and thus allowing the author to focus on the "rivalry" aspect.

Reinganum makes the following assumptions on the probability of discovery. Let $z_i(t)$ denote the knowledge stock of firm i at time t. For simplicity, assume that $z_i(t)$ is simply equal to firm i's accumulated research efforts:

$$\dot{z}_i(t) = u_i(t)$$

Let t_i denote the time at which firm i makes the discovery. It is a random variable. Assume the following cumulative distribution

$$\Pr(t_i < t) = 1 - \exp[-\lambda z_i(t)] \equiv F_i(t)$$

The probability that firm i has not made a discovery at time t is

$$\Pr(t_i > t) = \exp[-\lambda z_i(t)] = 1 - F_i(t)$$

The function $F_i(t)$ gives rises to the density function

$$f_i(t) = \frac{dF_i(t)}{dt} = \left[\lambda\frac{dz_i(t)}{dt}\right]\exp[-\lambda z_i(t)]$$

Let $Z(t)$ denote the sum of knowledge stocks of the n firms at time t:

$$Z(t) \equiv \sum_{k=1}^{n} z_k(t)$$

and

$$Z_{-i}(t) \equiv \sum_{k\neq i} z_k(t)$$

Assuming independence, the probability that no firm has made a discovery at t is

$$\Pr[t_i > t,\ i = 1, 2, ..n] = \Pi_{i=1}^{n}[1 - F_i(t)] = \exp[-\lambda Z(t)]$$

The probability density that firm i makes the discovery at time t while other firms remain unsuccessful at t is

$$
\begin{aligned}
f_i(t)\Pi_{j\neq 1}^{n}[1 - F_j(t)] &= \lambda u_i(t)\exp[-\lambda z_i(t)]\exp[-\lambda Z_{-i}(t)] \\
&= \lambda u_i \exp[-\lambda Z(t)] \qquad\qquad (5.10)
\end{aligned}
$$

Assume a common time horizon T (exogenously fixed). The expected payoff of firm i is the expected revenue minus the expected cost

$$
\begin{aligned}
J_i &= \int_0^T f_i(t)\Pi_{j\neq 1}^{n}[1 - F_j(t)]\Omega(t)e^{-rt}dt \\
&\quad - \int_0^T \Pr[t_i > t,\ i = 1, 2, \ldots, n]\frac{u_i^2}{2}e^{-rt}dt \\
&= \int_0^T e^{-rt}\left\{\lambda u_i\Omega(t) - \frac{u_i^2}{2}\right\}\exp[-\lambda Z(t)]dt
\end{aligned}
$$

Notice that because of the assumption that the distribution functions are exponential, only the aggregate knowledge matters. The relevant state variable is Z, not the individual z_i. This serves to simplify the analysis.

Open-loop Nash equilibrium

When firms use open-loop strategies, each firm i takes the time path of $u_k(t)$ ($k \neq i$) as given, and chooses the control variable $u_i(t)$ to maximize

J_i subject to

$$\dot{Z}_i(t) = u_i(t) + \sum_{j \neq i} u_j(t)$$

Reinganum assumes that

$$\Omega(t)e^{-rt} = W(t) = W \text{ (a positive constant)}$$

It is easy to show that, if $n \geq 2$, the OLNE efforts are

$$u_i(t) = \frac{2\lambda W(n-1)\exp[rt]}{(2n-1) - \exp[\lambda^2 W(e^{rt} - e^{rT})/r]} \tag{5.11}$$

which is an increasing function of t. (The numerator is increasing in t and the denominator is decreasing in t, so $u_i(t)$ is *increasing* in t.)

For a simple derivation of the OLNE of this game, see Dockner *et al.* (2000). Notice that the open-loop equilibrium strategy (5.11) is independent of the initial value of the state variable. This unusual feature is due to the assumption that the distribution function is exponential.

Markov-perfect Nash equilibrium

Each player i assumes that player $j(j \neq i)$ uses a strategy of the form

$$u_j(t) = \phi_i^{\text{FB}}(Z(t), t)$$

Reinganum (1982a) found that for this model, the MPNE strategies turn out to be degenerate, that is, they do not contain the state variable as an argument. In fact, she showed that the MPNE strategies for this model are identical to the OLNE strategies. This result can be shown to hold for a class of model where the state variables appear as an exponential term. Dockner *et al.* (2000) show that Reinganum's class of models is equivalent to a class of models that are linear in the state variables (in a well-defined sense), and it is known that for this latter class, OLNE and MPNE are identical. To see this, let us define a new state variable y using the transformation

$$y \equiv \exp[-\lambda Z(t)]$$

Then

$$\dot{y} = -\lambda y \sum_i u_i$$

We can now show that the MPNE strategies for the R&D race are independent of y. Let us suppose that

$$u_j(t) = \phi^{\mathrm{FB}}(y(t), t)$$

Then

$$\dot{y} = -\lambda y[u_i + (n-1)\phi^{\mathrm{FB}}(y(t), t)]$$

The HJB equation for player i is

$$-V_t^i = \max_{u_i} \left\{ \lambda u_i W(t)y - \frac{u_i^2}{2}ye^{-rt} - \lambda y V_y^i[u_i + (n-1)\phi^{\mathrm{FB}}(y, t)] \right\}$$

with the terminal condition $V^i(T, y(T)) = 0$.

Maximizing the RHS with respect to u_i gives the first-order condition

$$u_i = e^{rt}\lambda[W(t) - V_y^i]$$

Imposing symmetry, we get

$$\phi^{\mathrm{FB}}(y, t) = e^{rt}\lambda[W(t) - V_y^i]$$

Hence, the HJB equation becomes

$$-V_t^i = y\lambda^2 e^{rt} \left[\frac{W(t) - V_y^i}{2} \right] (W(t) - (2n-1)V_y^i)$$

Let us guess that player i's value function takes the form

$$V^i = b(t)y$$

It follows that the HJB equation becomes

$$b'(t) = -\frac{e^{rt}\lambda^2}{2}[b^2(t)(2n-1) - 2nW(t)b(t) + W^2(t)], \quad b(T) = 0$$

In the special case where W is a constant, one can solve this differential equation as follows.

Rewrite it as

$$\frac{db}{b^2(t) - \beta(t)b(t) + \gamma(t)} = -\alpha e^{rt} dt$$

where

$$\alpha = \frac{\lambda^2(2n-1)}{2}$$

$$\beta = \frac{\lambda^2 n W(t)}{\alpha}$$

$$\gamma = \frac{\lambda^2}{2\alpha W^2(t)}$$

Factorize $b^2(t) - \beta(t)b(t) + \gamma(t)$ into $(b(t) - b_1(t))(b(t) - b_2(t))$ where

$$b_1(t) = W(t)$$

$$b_2(t) = \frac{W(t)}{2n-1}$$

Note that

$$b_2(t) - b_1(t) = \frac{2W(t)(1-n)}{2n-1}$$

If $n - 1 \neq 0$, then $b_1(t)$ and $b_2(t)$ are distinct, and we can write the differential equation as

$$\frac{db}{b_2 - b_1}\left\{\frac{1}{b - b_2} - \frac{1}{b - b_1}\right\} = -\alpha e^{rt}dt \quad \text{if } n \neq 1$$

Now, assume that $W(t) = W$ contant, such that b_2 and b_1 are constants. This allows us to integrate both sides and get

$$\frac{1}{b_2 - b_1}\{\ln|b(t) - b_2| - \ln|b(t) - b_1|\} = \frac{-\alpha e^{rt}}{r} + A$$

where A is the constant of integration. Evaluating the above equation at $t = T$, making use of the boundary condition $b(T) = 0$, we can determine the constant A

$$A = \frac{\alpha}{r}e^{rT} + \frac{1}{b_2 - b_1}\ln\left(\frac{b_2}{b_1}\right) = \frac{\alpha}{r}e^{rT} + \frac{2n-1}{2W(1-n)}\ln\left(\frac{1}{2n-1}\right)$$

Thus

$$\frac{1}{b_2 - b_1} \ln \left[\frac{b(t) - b_2}{b(t) - b_1} \right] = \frac{1}{b_2 - b_1} \ln \left[\frac{1}{2n - 1} \right] + \frac{\alpha}{r}[e^{rT} - e^{rt}]$$

$$\ln \left[\frac{b(t) - b_2}{b(t) - b_1} \right] = \ln \left[\frac{1}{2n - 1} \right] - \frac{2\alpha}{r(2n - 1)}[e^{rT} - e^{rt}](n - 1)W$$

$$= \ln \left[\frac{1}{2n - 1} \right] - \lambda^2[e^{rT} - e^{rt}](n - 1)W/r$$

Define

$$\Delta = \lambda^2[e^{rT} - e^{rt}](n - 1)W/r$$

Applying the exponential operator, we get

$$\exp \left\{ \ln \left[\frac{b(t) - b_2}{b(t) - b_1} \right] \right\} = \exp \left\{ \ln \left[\frac{1}{2n - 1} \right] - \Delta \right\}$$

Hence

$$\frac{b(t) - b_2}{b(t) - b_1} = \frac{e^{-\Delta}}{2n - 1} = \frac{1}{(2n - 1)e^{\Delta}}$$

or

$$b(t) = \frac{W(1 - e^{\Delta})}{1 - (2n - 1)e^{\Delta}}$$

Thus, the value function is

$$V(t, y) = \frac{yW(1 - e^{\Delta})}{1 - (2n - 1)e^{\Delta}}$$

We can now solve for the feedback control rule, if $n \neq 1$:

$$u_i = e^{rt}\lambda[W(t) - V_y^i] = \frac{2(n - 1)\lambda W e^{rt} e^{\Delta}}{(2n - 1)e^{\Delta} - 1}$$

$$u_i = \frac{2(n - 1)\lambda W e^{rt}}{(2n - 1) - e^{-\Delta}}$$

$$= \frac{2(n - 1)\lambda W e^{rt}}{(2n - 1) - \exp\{\lambda^2[e^{rt} - e^{rT}](n - 1)W/r\}}$$

It follows that the feeback strategies are degenerate, that is, the MPNE and the OLNE coincide.

R&D with hazard rate uncertainty

Choi (1991) pointed out that Reinganum's assumption of a constant and known $\lambda > 0$ implies that a discovery is "technically possible", in the sense that the probability of discovery by date τ is arbitrarily close to 1 if τ is arbitrarily large. Choi assumes that λ is uncertain, but makes the simplifying assumption that R&D efforts are constant over time.

Malueg and Tsutsui (1997) consider the case where the true λ is not known, and decision makers think that either $\lambda = 0$ or $\lambda = \lambda^* > 0$. They assume that decision makers assign the prior probabilities of p_0 and $1 - p_0$ to these two events, and that p_0 gets revised upward as time proceeds and a discovery has not been made.

Let the event that $\lambda = 0$ be denoted by E_0 and the event that $\lambda = \lambda^*$ be denoted by E_1. Note that $\Pr(E_0 \cup E_1) = 1$ by assumption. Let $D_i(t)$ be the event that firm i discovers at some point during the small time interval $(t - \varepsilon, t)$ or before it, while all other firms remain unsuccessful during that interval.

Then, with the use of Bayes' rule

$$
\begin{aligned}
\Pr(D_i(t)) &= \Pr[D_i(t) \cap E_0] + \Pr[D_i(t) \cap E_1] \\
&= \Pr(E_0)\Pr[D_i(t) \mid E_0] + \Pr(E_1)\Pr[D_i(t) \mid E_1] \\
&= \Pr(E_1)\Pr[D_i(t) \mid E_1]
\end{aligned}
$$

Assume the exponential distribution for the discovery process. Taking the limit $\varepsilon \to 0$, we get, using (5.10) the probability density that firm i discovers at time t and other firms are by then still unsuccessful:

$$
\begin{aligned}
\Pr(E_1)\left\{\lim_{\varepsilon \to 0}\frac{1}{\varepsilon}\Pr[D_i(t) \mid E_1]\right\} &= (1 - p_0)f_i(t)\Pi^n_{j \neq 1}[1 - F_j(t)] \\
&= (1 - p_0)\lambda u_i \exp[-\lambda^* Z(t)]
\end{aligned}
$$

Next, the event that no firm makes a discovery before t is denoted by $\widetilde{D}(t)$. Then, with the use of Bayes' rule

$$
\begin{aligned}
\Pr(\widetilde{D}(t)) &= \Pr[\widetilde{D}(t) \cap E_0] + \Pr[\widetilde{D}(t) \cap E_1] \\
&= \Pr(E_0)\Pr[\widetilde{D}(t) \mid E_0] + \Pr(E_1)\Pr[\widetilde{D}(t) \mid E_1] \\
&= \Pr(E_0) + \Pr(E_1)\Pr[\widetilde{D}(t) \mid E_1] \\
&= p_0 + (1 - p_0)\exp\left[-\lambda^* Z(t)\right]
\end{aligned}
$$

The expected payoff of firm i is then

$$J^i = \int_0^T e^{-rt} \Omega(t)(1 - p_0)\lambda u_i \exp[-\lambda^* Z(t)]dt$$

$$- \int_0^T e^{-rt} \frac{u_i^2}{2} \{p_0 + (1 - p_0) \exp[-\lambda^* Z(t)]\}dt \qquad (5.12)$$

Now, define

$$y(t) = \exp[-\lambda^* Z(t)]$$

Let us denote by $p(t)$ the posterior probability that $\lambda = 0$, given that no firm has made a discovery by time t. We obtain $p(t)$ using Bayes' rule:

$$p(t) \equiv \Pr[\lambda = 0 \mid t_i < t \text{ for all } i] \equiv \Pr[E_0 \mid \widetilde{D}(t)] = \frac{\Pr[\widetilde{D}(t) \cap E_0]}{\Pr(\widetilde{D}(t))}$$

$$= \frac{\Pr[\widetilde{D}(t) \mid E_0] \Pr(E_0)}{\Pr(\widetilde{D}(t))} = \frac{1.p_0}{p_0 + (1 - p_0)y(t)} \qquad (5.13)$$

This gives us the relationship

$$p_0 + (1 - p_0)y(t) = \frac{p_0}{p(t)} \qquad (5.14)$$

Hence

$$(1 - p_0)y(t) = \frac{p_0(1 - p(t))}{p(t)} \qquad (5.15)$$

From Eq. (5.13)

$$\dot{p}(t) = \frac{-p_0(1 - p_0)\dot{y}(t)}{[p_0 + (1 - p_0)y(t)]^2} = -p_0(1 - p_0)\dot{y}(t) \left[\frac{p(t)}{p_0}\right]^2$$

$$= \frac{[p(t)]^2(1 - p_0)y(t)\lambda^* \sum u_i}{p_0} \qquad (5.16)$$

Substitute Eq. (5.15) into Eq. (5.16) to obtain

$$\dot{p} = p(t)[1 - p(t)]\lambda^* \sum u_i(t)$$

This equation shows that if $u_i > 0$, then $p(t)$ will converge to 1 as $t \to \infty$.

The payoff function (5.12) can be rewritten by noting that from Eq. (5.14)

$$p_0 + (1 - p_0) \exp[-\lambda^* Z(t)] = \frac{p_0}{p(t)}$$

and that, using Eq. (5.15)

$$e^{-rt}\Omega(t)(1-p_0)\lambda^* u_i \exp[-\lambda^* Z(t)] = W(t)[\lambda^* u_i](1-p_0)y(t)$$
$$= W(t)[\lambda^* u_i]\frac{p_0(1-p(t))}{p(t)}$$

Hence, the optimal control problem facing firm i is to choose $u_i(t)$ to maximize

$$\int_0^T W(t)[\lambda^* u_i]\frac{p_0(1-p(t))}{p(t)}dt - \int_0^T e^{-rt}\frac{p_0}{p(t)}\left[\frac{u_i^2(t)}{2}\right]dt$$

subject to

$$\dot{p} = p(t)[1-p(t)]\lambda^* \sum u_i(t)$$

This differential game does not seem to have a closed form solution. MPNE strategies can be obtained numerically, see Malueg and Tsutsui (1997). They found that a firm may drop out of the R&D race if it has tried long enough without success.

5.3.2. *Technology-adoption games*

The games of technology adoption that we survey in this subsection are dynamic games without the usual transition equation. In the simplest technology-adoption model, the relevant state of the system is the identity of the firms that have switched over to a new technology.

A simple technology-adoption game: equilibrium with precommitment

Reinganum (1981b) considers two *ex ante* identical firms, which we denote by a and b. Let T_j be firm j's technology adoption time ($j = a, b$). Assume the cost of adoption is lump-sum $K(T_j) > 0$ (in present value terms). When both firms use the old technology, they each earn π_0 per unit of time. When both use the new technology, they each earn π_N. When one firm uses the old technology and the other uses the new technology, they earn respectively π_ℓ and π_h. The following assumptions are made concerning the relationship among these different profit levels.

Assumption TA1

$$\pi_h > \pi_N > \pi_\ell \quad \text{and} \quad \pi_h > \pi_0 > \pi_\ell$$

Assumption TA2

$$(\pi_h - \pi_0) - (\pi_N - \pi_\ell) = \alpha > 0$$

that is, the gain when you are the first one to innovate is greater than the gain when you are the second to innovate.

Clearly assumptions TA1 and TA2 are satisfied if the demand function is linear and firms produce under constant returns to scale.

Reinganum makes an important behavioral assumption: firms adopt open-loop (i.e., precommitment) strategies. Each chooses an adoption date, taking the other's adoption date as given. (It is as if at the beginning of the game, each firm makes a binding announcement specifying its date of technology adoption, knowing the date of adoption of the other firm.)

Let $D(t) = \exp(-rt)$ where r is rate of interest. If firm a is the first to innovate, the present value of its profit stream net of innovation (adoption) cost is

$$g^1(T_a, T_b) = \int_0^{T_a} \pi_0 D(t)dt + \int_{T_a}^{T_b} \pi_h D(t)dt + \int_{T_b}^\infty \pi_N D(t)dt - K(T_a)$$

Note that $g^1(T_a, T_b)$ is increasing in T_b because $\pi_H > \pi_N$.

If firm a is the second one to innovate, that is, $T_a > T_b$, the present value of its profit streams net of innovation (adoption) cost is

$$g^2(T_a, T_b) = \int_0^{T_b} \pi_0 D(t)dt + \int_{T_b}^{T_a} \pi_\ell D(t)dt + \int_{T_a}^\infty \pi_N D(t)dt - K(T_a)$$

Note that $g^2(T_a, T_b)$ is increasing in T_b because $\pi_0 > \pi_L$.

Given the choice T_a and T_b, the payoff to firm a is

$$V_a(T_a, T_b) = \begin{cases} g^1(T_a, T_b) & \text{if } T_a \leq T_b \\ g^2(T_a, T_b) & \text{if } T_a > T_b \end{cases}$$

Thus, $V_a(T_a, T_b)$ is increasing in T_b. That is, *an increase in T_b always benefits firm a.*

Given T_b, suppose firm 1 considers the merit of being the first innovator. Its marginal net gain of delaying T_a a bit (but T_a is still earlier than T_b) is

$$\frac{\partial g^1(T_a, T_b)}{\partial T_a} = [-K'(T_a)] - \{(\pi_h - \pi_0)D(T_a)\} \tag{5.17}$$

The term inside the curly brackets is the marginal cost of the forgone increase in profit flow. This marginal cost is positive by TA2, but it falls as

T_a increases. The term inside the square bracket is the marginal benefit of delaying the adoption cost. If we write $K(t) = e^{-rt}C(t)$, then

$$[-K'(T_a)] = e^{-rT_a}(rC(T_a) - C'(T_a))$$

and $[-K'(T_a)] > 0$ if the cost $C(t)$ falls at a rate greater than the interest rate.

The second derivative is

$$\frac{\partial^2 g^1(T_a, T_b)}{(\partial T_a)^2} = [-K''(T_a)] + r(\pi_h - \pi_0)D(T_a)$$

Assumption TA3 (i) Adoption cost initially falls fast enough, such that

$$[-K''(t)] + r(\pi_h - \pi_0)D(t) < 0 \quad \text{for all } t > 0$$

and (ii) the following equation has a positive solution $T_a^* < \infty$:

$$[-K'(T_a^*)] - (\pi_h - \pi_0)D(T_a^*) = 0$$

The first part of assumption TA3 implies that $g^1(T_a, T_b)$ is strictly concave in T_a for $T_a < T_b$. The marginal net gain for the second mover if it delays T_a a bit is

$$\frac{\partial g^2(T_a, T_b)}{\partial T_a} = [-K'(T_a)] - (\pi_N - \pi_\ell)D(T_b) \qquad (5.18)$$

By Assumption TA2, if we evaluate the expressions (5.17) and (5.18) at $T_a = T_b$, we see that the right-hand derivative is greater than the left-hand derivative.

Reinganum's main result is that under the assumption of precommitment (the equivalent of open-loop strategy) the Nash equilibrium will exhibit the property of "technology diffusion": in equilibrium, one firm will innovate first, and the other will innovate at a later date.

Fudenberg and Tirole (1985) supply an elegant proof that in the *precommitment equilibrium, the first innovator earns a higher lifetime payoff.* The proof is as follows. Let (T_a, T_b) denote the adoption times of (a, b). Consider a Nash equilibrium (T_a^*, T_b^*). Suppose without loss of generality that $T_a^* < T_b^*$. Then

$$V_a(T_a^*, T_b^*) \geq V_a(T_b^*, T_b^*) = V_b(T_b^*, T_b^*) > V_b(T_a^*, T_b^*)$$

The first inequality comes from the definition that T_a^* is a best reply of a to T_b^*. The equality comes from the symmetry of the payoff functions. The final strict inequality comes from the fact that each firm's payoff goes up if its rival's adoption date is moved to a later date.

It is easy to see that T_a^* satisfies the first-order condition

$$[-K'(T_a^*)] - \{(\pi_h - \pi_0)D(T_a^*)\} = 0$$

Given Assumption TA3, this first-order condition identifies a unique T_a^*. Similarly, T_b^* satisfies the following first-order condition

$$[-K'(T_b^*)] - \{(\pi_N - \pi_\ell)D(T_b^*)\} = 0$$

Technology adoption without precommitment: pre-emption equilibrium with payoff equalization

Fudenberg and Tirole (1985) point out that the first mover advantage found in the pre-commitment equilibrium of Reiganum's model is problematic: why does not the second mover try to preempt its rival and become the first mover? In fact, if firm b innovates just an instant before T_a^*, then firm a will no longer want to stick to T_a^*. Fudenberg and Tirole seek to characterize MPNE. In such equilibria, firms use feedback strategies, such that equilibrium strategies must remain equilibrium in every subgame. Recall that a feedback strategy means that the player's optimal action (best response) is dependent on the "observed state" of the system. In the technology adoption model, the state at date t can be denoted by the vector $(S_a(t), S_b(t))$ where $S_i(t)$ can take on either of the two discrete values, 0 or 1, where $S_i(t) = 0$ means that firm i is still using the old technology prior to t, and $S_i(t) = 1$ means that firm i has been using the new technology over some time interval $(t-\varepsilon, t)$ for some $\varepsilon > 0$. (There is a bit of a problem about timing here: if firm i adopts the new technology at t, we may denote this by $S_i(t) = 0$ and $S_i(t+\varepsilon) = 1$ for all $\varepsilon > 0$.)

Fudenberg and Tirole obtain an important result, namely the dissipation of first-innovator advantage. The potential first-innovator advantage stimulates preemption up to the point of time where the gain from being the first innovator is just offset by the extra cost of early adoption. They also find that pre-commitment equilibrium Pareto dominates pre-emption equilibrium.

Fudenberg and Tirole's construction of the pre-emption equilibrium relies on the the construction of a function called the leader's payoff function. Define T_2^* as the solution of the equation

$$[-K'(T)] - (\pi_N - \pi_\ell)D(T) = 0$$

Suppose a firm manages to innovate at some time t before the other firm has a chance to move. Then the latter becomes the "follower" and it would choose T_2^* if $t \leq T_2^*$. If for some reason the innovation time of the first innovator exceeds T_2^* (i.e., $t > T_2^*$) then, because of the assumption that the payoff function is strictly quasi-concave in one's own innovation date, the follower will innovate "immediately after" t, that is, at t as well. So, the follower's payoff, expressed as a function of the date of adoption t by the first innovator, is

$$F(t) = \begin{cases} V(T_2^*, t) & \text{if } t < T_2^* \\ V(t, t) & \text{if } t \geq T_2^* \end{cases}$$

The leader's payoff is then

$$L(t) = \begin{cases} V(t, T_2^*) & \text{if } t < T_2^* \\ V(t, t) & \text{if } t \geq T_2^* \end{cases}$$

Define

$$M(t) \equiv V(t, t)$$

$M(t)$ is the payoff to each firm if both firms innovate at the same time. It can be shown that $M(t)$ reaches its unique maximum at some $\widehat{T}_2 > T_2^*$.

Define T_1^* as the solution of the equation

$$[-K'(T)] - (\pi_h - \pi_0)D(T) = 0$$

There are two cases.

Case A: $L(T_1^*) > M(\widehat{T}_2)$

Case B: $L(T_1^*) \leq M(\widehat{T}_2)$

In Case A, each firm would want to innovate at T_1^* if it can be sure that the other firm will innovate at a later date. But if a firm knows that the other firm's strategy is to adopt (*without precommitment*) at T_1^*, it will preempt at time $T_1^* - \varepsilon$, because it knows that if this preemption is successful, the other firm will not innovate at T_1^* as previously planned, but will choose T_2^* instead. But the other firm, knowing this preempting attempt, will want to preempt its rival at an earlier date (to avoid being preempted). This process continues until some time $T_1 < T_1^*$ is reached where the payoff obtained by preempting is as low as the payoff of being a follower (who would innovate at T_2^*.) This is then the only equilibrium (unique, up to a relabelling of firms). In this pre-empting equilibrium, there is diffusion and the payoffs of the two firms are *equalized*.

In Case B, there are two classes of equilibria. The first class is the (T_1, T_2^*) diffusion equilibrium. To describe the second class, let $S \leq \widehat{T}_2$ be the time such that $M(S) = L(T_1^*)$. Then any simultaneous adoption date $t \in [S, \widehat{T}_2]$ (with "adopt" in every subgame from t on) is a perfect equilibrium.

All simultaneous-adoption equilibria Pareto dominate the pre-commitment equilibrium (T_1^*, T_2^*), which in turn dominates the perfect diffusion equilibrium (T_1, T_2^*).

Technology adoption: a waiting game among heterogeneous firms

Katz and Shapiro (1987) modify the Fudenberg–Tirole model to study a waiting game among heterogeneous firms (in an asymmetric nonstationary, complete information setting). The difference is that Katz and Shapiro specify that "adoption" here means developing a new patented production process, and hence once a firm "adopts", other firms cannot adopt the same technology. Either the noninnovators drop out, or they imitate the innovator's technology or buy a license to use it. Katz and Shapiro show that the losers may wish to lose as soon as possible, since the innovating firm is providing a public good to the industry. They also show that under certain conditions, there exists an equilibrium without innovation, and another equilibrium which is Pareto inferior, where one firm innovates "in self-defense" for fear that the other firm would do so. In this model, the potential second-mover advantage may yield subgame perfect equilibria in which preemption and payoff equalization do not arise.

Technology adoption when better technology is available later

Dutta *et al.* (1995) developed a model of *continuing* innovations and improvements where the players are *ex ante* identical firms. Unlike the Fudenberg–Tirole model where firms dissipate all intramarginal rents (that would go to the first adoptor under precommitment) by the motive to preempt, in the DLR model firms can improve the technology by waiting (the quality level of a firm's *potential* product improves at a constant rate, until adoption takes place). Let $R(x)$ denote the flow monopoly profit of a monopolist whose product attains the quality level x. If a firm i has introduced a good at quality level x_i and later firm j introduces a competing product at quality level x_j, the duopolists will earn $\pi_i(x_i, x_j)$

and $\pi_j(x_j, x_i)$. The authors assume that prior to adoption, x_i and x_j grow at a common constant rate. Once a firm i has adopted, its x_i stops growing. The authors also use the concept of subgame perfection. A pure strategy of firm i, denoted by s_i, is a function that specifies at any time t a decision to "adopt" or "not adopt" if the firm has not already adopted. This decision is conditioned on the history H_t which records if j has adopted prior to t, and if so, at what time. A strategy pair (s_1^*, s_2^*) is a subgame-perfect equilibrium if for all history H_t the strategy s_i^* maximizes player i's payoff in the continuation game, given s_j^*.

They show the existence of an equilibrium (called *maturation equilibrium*) where there is *"rent escalation"*: a later entry yields a higher lifetime profit. The other type of equilibrium (preemption, with rent dissipation) may also exist. In one example, they show that maturation equilibria exist if and only if there is a strong diversity of preferences.

In the maturation equilibrium, each firm would prefer to be the late innovator. An incumbent would delay an innovation in order not to cannibalize an existing product. This may give the "nonincumbent" the ability to commit to be the later innovator.

This model relies on a very strong assumption, namely firm i's duopoly profit is a function of the difference $x_i - x_j$.

Technology adoption without the single-peakedness assumption

In the Fudenberg–Tirole game, the flow revenues of the two firms at any time t depend only on the current "technology status" of the two firms, that is, whether they both use the old technology, or both use the new technology, or one uses the old one while the other uses the new one. Thus a firm's revenue flow at any time depends only on its technology status and that of the rival, and is *independent of innovation dates*. This is quite "unrealistic" for innovation games. It would seem more plausible to assume that the later is the innovation date, the lower is the operating cost, or the higher is the quality of the good. This more realistic assumption is adopted by Hoppe and Lehman-Grube (2005).

Hoppe and Lehman-Grube also notice that the earlier models assume that the first mover's payoff is single-peaked in the first mover's innovation time. When this assumption is relaxed, a new method of analysis must be found. Hoppe and Lehman-Grube propose a new approach for a simple timing game which does not require single-peakedness.

A firm's lifetime payoff is denoted by $V_a(t_a, t_b)$ (for firm a) or $V_b(t_b, t_a)$ (for firm b). If firm i chooses t_1 and firm j chooses t_2 with $t_2 > t_1$, we call firm i the leader, and the other firm the follower. They assume that there exist two piece-wise continuous functions g_1 and g_2 defined over the set $\{(t_1, t_2) \text{ st } 0 \le t_1 \le t_2\}$ such that $g_1(t_1, t_2) = g_2(t_1, t_2)$ if $t_1 = t_2$, and

$$V_a(t_a, t_b) \begin{cases} g_1(t_a, t_b) & \text{if } t_a < t_b \\ g_2(t_b, t_a) & \text{if } t_a \ge t_b \end{cases}$$

$$V_b(t_b, t_a) \begin{cases} g_1(t_b, t_a) & \text{if } t_b < t_a \\ g_2(t_a, t_b) & \text{if } t_b \ge t_a \end{cases}$$

Following Simon and Stinchcombe (1989), the continuous time formulation is taken as the limit of discrete time "with a grid that is infinitely fine". The model has two possible interpretations: (a) process innovation timing, and (b) product innovation timing.

Hoppe and Lehman-Grube assume the firm that innovates first enters the market and becomes a monopolist. When the second firm innovates, the industry becomes a duopoly. If one firm innovates first, it is the monopolist until the second firm innovates. It follows that

$$g_1(t_1, t_2) = \int_{t_1}^{t_2} R_m(t_1) D(t) d\tau + \int_{t_2}^{\infty} R_1(t_1, t_2) D(t) d\tau$$

$$- K(t_1) \quad \text{with } t_1 \le t_2$$

$$g_2(t_1, t_2) = \int_{t_2}^{\infty} R_2(t_1, t_2) D(t) d\tau - K(t_2) \quad \text{with } t_1 \le t_2$$

Here $K(t_1)$ is the cumulative cost of R&D up to time t_i.

(A) **Process innovation timing** Assume the product is of fixed quality and characteristics. The two firms, if they are both active, produce the same homogenous good. A firm decides to enter (to produce) if it thinks its unit cost has fallen far enough. Assume the unit cost falls at a constant proportional rate: $c_i = e^{-\alpha t}$. Once a firm enters, its unit cost stops falling. Assume the demand function $p = 1 - q$. The monopoly profit of the first entrant (with cost c_1) is

$$\widehat{R}_m = \frac{1}{4}(1 - c_1)^2$$

When firm 2 enters (of course, $c_2 < c_1$), either $c_2 \ge 2c_1 - 1$, or $c_2 < 2c_1 - 1$. The first case is called the non-drastic innovation, because the existing firm can still produce without incurring losses. In the second case, called the "drastic innovation", the incumbent shuts down, and the second innovator becomes a monopolist. Thus, the revenue flow is

$$\widehat{R}_1 = \frac{1}{4}(1 - 2c_1 + c_2)^2 \text{ and } \widehat{R}_2 = \frac{1}{4}(1 - 2c_2 + c_1)^2 \quad \text{if } c_2 \geq 2c_1 - 1$$

$$\widehat{R}_1 = 0 \text{ and } \widehat{R}_2 = \frac{1}{4}(1 - c_2)^2 \quad \text{if } c_2 < 2c_1 - 1$$

(B) **Product innovation timing** Assume that variable production cost is constant, say zero, but the quality of a firm's potential product falls over time (as along as it has not started producing it). The product quality s is a function of time. Assume $\dot{s}(t) = 1$ and $s(0) = 0$, so that $s = t$. (product quality improves "one step at a time"). R&D cost per period (before adoption of the product) is $K = \lambda s$. Consumers are indexed by $\theta \in [0, 1]$. Each consumer buys one unit of the product. The net surplus of consumer θ if he buys one unit of quality s_i at the price p_i is $U = s_i \theta - p_i$. Then

$$R_m = \frac{1}{4}t_1$$

$$R_1 = t_1 t_2 \frac{t_2 - t_1}{(4t_2 - t_1)^2} \quad \text{and} \quad R_2 = 4t_2^2 \frac{t_2 - t_1}{(4t_2 - t_1)^2}$$

Assume for the moment that if the leader innovates at time t then the follower will react in a way that can be summarized by a reaction function $\phi(t)$, where $\phi(t) > t$. Then, the leader's lifetime payoff if it innovates at t is

$$L(t) = g_1(t, \phi(t))$$

and the follower's lifetime payoff is

$$F(t) = g_2(t, \phi(t))$$

Assume if the follower is indifferent between two dates at which it may innovate, it will choose the earlier one. In general, the curve $L(t)$ can have multiple peaks. It turns out that that any "multiple-peaked $L(t)$ curve" can be replaced by its "envelope" which is by definition quasi-concave.

Under certain assumptions, the equilibrium is unique and, depending on parameter values, the follower's equilibrium payoff may be equal to or greater than the leader's equilibrium payoff.

Chapter 6

DYNAMIC GAMES IN PUBLIC ECONOMICS

This chapter surveys dynamic game models in public economics. In Section 6.1, we review a number of models of private contributions to a public good. These models mainly consider a stock public good (as distinct from a flow public good) that can be accumulated. An exception is the model of Benchekroun and Long (2008a) where the state variable is not a stock of public good, but is simply an indicator of the stock of mutual trust in a community, which can be built up by agents who play a non-cooperative game. This intangible stock can help agents to increase their contributions to a flow public good, period after period, thus achieving higher welfare than the static Nash equilibrium welfare level. In Section 6.2, we turn to a model of voluntary contributions to a discrete public good: the public good can be built only after the total accumulated contribution equals a certain level. Section 6.3 turns to the design of tax and subsidy rules to induce a far-sighted monopolist (or oligopolists) to achieve the social optimum. This problem may be regarded as a leader–follower game: the regulator is a leader, and the firms are followers. Section 6.4 surveys dynamic models of redistributive taxations among factor owners (with the main focus on the taxing of capital income). Game-theoretic issues relating to intergenerational equity and parental altruism are discussed in Section 6.5. Finally, in Section 6.6 we survey game-theoretic models of fiscal competition and electoral incentives set in a dynamic context.

6.1. Contributions to a Stock of Public Good

6.1.1. *Tangible public goods*

The static theory of voluntary contributions to a public good (developed by Warr, 1983; Bergstrom *et al.*, 1986; Boadway *et al.*, 1989a,b), has been extended, with the use of dynamic game theory, to the case where the public good is a *stock* that can be built up over time. The main contributors to this literature are Fershtman and Nitzan (1991), Wirl (1996), and Itaya and Shimomura (2001).[1] Examples of private contributions to a public

---

[1]For an excellent exposition of the static theory of public goods and externalities, see Cornes and Sandler (1996).

good include the voluntary maintenance of dykes, donations to charity organizations, and private efforts in the monitoring and denouncing of corrupt behavior of government officials. As an example of a stock of public good, consider a public library in a local community whose members make donations for the purchase of books. The stock of books increases over time if additional contributions exceed the rate of replacement of damaged books. Another example is the efforts of national governments to cut the emissions of GHGs. Murdoch and Sandler (1986) argue that military expenditures of NATO allies can to some extent be regarded as private contributions to a stock of public goods.

Modeling a dynamic game of private contributions to a divisible stock of public good, K, Fershtman and Nitzan (1991) show that in a feedback Nash equilibrium, individual contributions are strategic substitutes: if an individual contributes more, the stock will increase, which induces other individuals to reduce their contributions. Since individuals know this, they have weaker incentives to increase their contributions. This is in sharp contrast to OLNE, where each individual thinks that varying his time path of contributions does not affect the time paths chosen by others. As a consequence, the OLNE results in a higher steady-state stock of the public good.

Let us outline the model. Assume that each of the N players has a concave and quadratic benefit function, $f(K) = aK - bK^2$, with $a > 0$ and $b \geq 0$, and a quadratic effort cost function $C(u) = (1/2)u^2$. The transition equation is linear, $\dot{K}(t) = Nu - \delta K$, where δ is the rate of depreciation. The authors obtain Nash equilibrium feedback strategies that are linear in the state variable (the stock of public good). The long-run stock of the public good when agents play feedback strategies is shown to be smaller than that obtained when agents play open-loop strategies.[2] Under both the OLNE and the feedback equilibrium, agents "free ride" (they do not contribute as much as they would under co-operation), but the free riding problem is more severe under the feedback equilibrium.

It is a good exercise to find the feedback Nash equilibrium in the case of a convex benefit function, say $f(K) = \gamma K^2$, where $\gamma > 0$. One can show that if γ is sufficiently small, the Nash equilibrium feedback strategy for

---

[2]Note that in the special case where $b = 0$, so that the game is linear in the state variable, the OLNE and the feedback equilibrium are identical: u^* is a constant.

each player is of the form $u = \beta K$ where $N\beta < \delta$, that is, the stock of public good falls to zero asymptotically.[3]

The Fershtman–Nitzan model is a mirror image of the model of private contributions to a public bad (a stock of pollution) that Dockner and Long (1993) investigate in the transboundary pollution game (see Chap. 2). Recall that Dockner and Long (1993) show that in addition to linear feedback strategies, there are also pairs of non-linear strategies that constitute MPNE.[4] Not surprisingly, this property applies also to the game of contributions to a public good. This is confirmed by Wirl (1996).

The issue of existence of non-linear strategies is further elucidated by Itaya and Shimomura (2001) who use the differential game approach to justify the static conjectural variations equilibrium that some authors obtain in static (or partially dynamic) games of private contributions. Itaya and Dasgupta (1995) develop a dynamic model of public good contribution, but they assume that contributors are not farsighted, maximizing instantaneous utility at each point of time. In that model, agents have heterogeneous initial conjectural variations. Itaya and Dasgupta (1995) show that under certain plausible assumptions the conjectures of agents converge in the long run to a consistent conjectural variations equilibrium. The model of Itaya and Shimomura (2001) is an attempt to rationalize conjectural variations by using a dynamic game where individuals optimize over time. This model is more general (in terms of functional forms) than Wirl (1996).[5]

Consider a community consisting of N agents. Agent i allocates his exogenous income flow $Y_i(t)$ between consumption $C_i(t)$ and contribution $X_i(t)$ to aggregate investment in a stock of public good $G(t)$, where

$$\dot{G}(t) = -\delta G(t) + \sum_{i=1}^{N} X_i(t)$$

---

[3]One would get a quadratic equation in β with two positive roots. Only the smaller positive root ensures that the transversality condition $\lim_{t \to \infty} e^{-rt} V(K(t)) = 0$ is satisfied, where $V(.)$ is the value function. See, for example, Chang (2004) and Dockner *et al.* (2000) on transversality conditions relating to value functions.

[4]The first proof of this type was offered by Tsutsui and Mino (1990) for a model of sticky price duopoly.

[5]Another direction of generalization is taken by Marx and Matthews (2000) who introduce imperfect information and focus on Baysesian equilibria of contributions to a public good.

Here, $\delta > 0$ is the rate of depreciation of the stock $G(t)$. The utility function of agent i is

$$u_i = U^i(C_i, G)$$

where $C_i(t) = Y_i(t) - X_i(t)$. His lifetime welfare is

$$W_i = \int_0^\infty e^{-rt} U^i(C_i(t), G(t)) dt$$

where r is the rate of discount.

If a social planner maximizes the sum $\sum W_i$, the steady-state stock of public good G_∞ must satisfy the following dynamic counterpart of the static Samuelsonian condition:

$$\sum_{i=1}^N \frac{U_G^i(C_i, G)}{U_C^i(C_i, G)} = \delta + r$$

That is, the sum of marginal rates of substitution must equal the long-run marginal rate of transformation.

In the absence of a central planner, consider the dynamic game among the N individuals. Suppose all individuals use feedback strategies, so that agent i's contribution X_i is a function of the stock, $X_i = \phi^i(G)$. Given an equilibrium profile of feedback strategies, one can show that the corresponding steady-state stock of public good G_∞^{FB} satisfies the following condition for each i:

$$\frac{U_G^i(C_i, G)}{U_C^i(C_i, G)} = \delta + r - \sum_{j \neq i} \phi_G^j(G) \tag{6.1}$$

where $\phi_G^j(G)$ denotes the derivative of $\phi^i(G)$. The last term on the RHS of Eq. (6.1) reflects the fact that individual i knows that if he contributes more, pushing G up, others will contribute less (if $\phi_G^j(G) < 0$), and this knowledge reduces his incentive to contribute. This term resembles the negative conjectural variations in static models.

Assume that the equilibrium strategy profile yields a steady state G_∞^{FB} that is stable, that is,

$$\frac{d\dot{G}}{dG} = -\delta + \sum_{i=1}^N \phi_G^i(G_\infty^{\mathrm{FB}}) < 0$$

One can then show that, if the marginal rate of substitution is diminishing, the steady state G_∞^{FB} under the feedback Nash equilibrium is below the level

G_∞ that the social planner would desire. A similar argument shows that this underprovision result also applies if agents use open-loop strategies.

Itaya and Shimomura (2001) specifically address the question of whether this model implies positive or negative conjectural variations at the steady state.[6] For this purpose, they assume identical individuals and specify that the utility function takes the separable form

$$U^i(C_i(t), G(t)) = u(C_i - \phi^i(G)) + h(G)$$

where $h(G)$ is the utility derived from the public good. Under this specification, one can show that in the symmetric feedback Nash equilibrium, the contribution strategy satisfies the following condition:

$$\phi'(G) = \frac{\frac{h'(G)}{u'(Y-\phi(G))} - r - \delta}{1 - N + (N\phi(G) - \delta G)\frac{u''(Y-\phi(G))}{u'(Y-\phi(G))}} \tag{6.2}$$

This equation yields a family of solutions, each different from others by the value taken by the constant of integration. At a steady state (with $G = \overline{G}$), Eq. (6.2) yields

$$\phi'(\overline{G}) = \frac{1}{N-1}\left[r + \delta - \frac{h'(\overline{G})}{u'(Y - \phi(\overline{G}))}\right] \tag{6.3}$$

At $\overline{G}$, it holds that $\delta\overline{G} = N\phi(\overline{G}) \equiv G_{-i}(\overline{G}) + \phi(\overline{G})$. This equation implies that steady-state conjectural variations must satisfy

$$\frac{dG_{-i}}{dX_i} = \frac{1 - \delta + (N-1)\phi'(\overline{G})}{\delta - (N-1)\phi'(\overline{G})} \tag{6.4}$$

Equations (6.3) and (6.4) yield the steady-state conjectural variations

$$\frac{dG_{-i}}{dX_i} = \frac{1 + \rho - \frac{h'(\overline{G})}{u'(Y-\phi(\overline{G}))}}{\frac{h'(\overline{G})}{u'(Y-\phi(\overline{G}))} - \rho}$$

Itaya and Shimomura (2001) show that the sign of conjectural variations can be positive or negative.[7]

---

[6]Dockner (1992) links conjectural variations to feedback Nash equilibrium in the context of a dynamic oligopoly. See also Dryskill and McCafferty (1989).

[7]The possibility of positive conjectural variations was indicated by Cornes and Sandler (1984). Sugden (1985) argued that positive conjectures in public good contributions imply irrational behavior or inferior private goods. Cornes and Sandler (1985, 1996)

Yanase (2003) takes a step further and asks what constant subsidy rate would achieve the socially optimal long-run stock G_∞. The answer depends on whether agents use open-loop strategies or feedback strategies. Assume that for each dollar an individual contributes, he/she gets back a fraction s_i through government subsidies. The total subsidy paid out by the government is $\sum_i s_i X_i$. Assuming that the government's budget is balanced by levying lump-sum taxes (denoted by L_i for individual i), the following condition holds:

$$\sum_i L_i = \sum_i s_i X_i$$

Yanase (2003) assumes that individuals take the lump-sum tax as given. He shows that if agents are identical and play open-loop strategies, to achieve efficiency at the steady state, the government should set the subsidy rate at $s^* = 1 - (1/N)$. This is the same as the result obtained in static models, for example, Boadway *et al.* (1989a).

When agents play feedback strategies, the (long-run) efficiency-inducing subsidies depend on whether agents use linear feedback strategies or non-linear ones.[8] Yanase (2003) shows that in the case of identical players using the same feedback rule $\phi(G)$, the (long-run) efficiency-inducing subsidy rate in the steady state is

$$s^* = 1 - \frac{r + \delta}{N[r + \rho - (N-1)\phi_G(G)]}$$

Notice that Yanase's subsidy does not replicate the optimal path of accumulation, it only achieves the long-run optimal stock.

6.1.2. *Trust: an intangible stock of public good that promotes co-operative behavior*

While the public good in the above-mentioned papers enters either the utility function or the production function, Benchekroun and Long (2008a) study a special type of stock public good that *affects neither utility nor production* directly, yet it is useful for society as a co-ordination device, helping a community to provide a monotone-increasing stream of flow public goods. Their stock public good is an intangible one: it is simply an

---

point to social norms and other non-market institutions as ways of eliciting positive conjectures.

[8]Recall that non-linear feedback strategies can arise even in linear-quadratic games, see for example, Tsutsui and Mino (1990), Dockner and Long (1993), Wirl (1996), and Itaya and Shimomura (2001).

indicator of the history of co-operation among individuals in a community. When this index is high, individuals trust each other more, and are more willing to provide community services (flow public goods). The idea that an abstract, non-tangible stock of public good can stimulate the voluntary contributions to tangible flow public goods seems appealing. Benchekroun and Long (2008a) find that there are infinitely many non-linear MPNE, some of which can approximate the efficient provision of flow public goods.

The theoretical model of Benchekroun and Long (2008a) shows how behavior based on trust can be self-sustaining: if Sue believes that John will "do good", it will be in Sue's best interest to "do good" and adopt a "co-operative" behavior. This is in sharp contrast with models where an individual performs a good act to induce other agents to follow. In the model presented by Benchekroun and Long (2008a), the causality is reversed: given other players' co-operative behavior, an agent finds it in his interest to be co-operative.[9]

Benchekroun and Long (2008a) rely neither on repeated games with trigger strategies which punish deviation by reverting to the static Nash equilibrium, nor on reputation building (Kreps, 1990). Instead, the authors model the conditioning of behavior on a social history of trust and co-operation. Such a history need not be the history of plays of current players. A history of behavior of past players (who have exited the game) may influence the behavior of new players.[10] Benchekroun and Long (2008a) postulate the existence of a summary measure of social history that may be called the "stock of co-operation". This stock increases when current agents behave co-operatively. It is also subject to decay (as people tend to forget events that happened a long time ago). The authors focus on the pure case where there are neither guilt nor penalties in the sense of trigger strategies.

From a control-theoretic perspective, the stock of co-operation is a state variable with a natural rate of depreciation. When non-cooperative agents use decision-rule strategies that dictate more co-operation when higher values of this state variable are observed, they gradually overcome the free-riding problem. An implication of the model is that two communities with identical resources, utility functions and production functions may end up with two different steady-state levels of public good, if in one

---

[9]Experiments by Kurzban and Houser (2004) show the majority of agent is of the reciprocal type. In Benchekroun and Long (2008a), however, reciprocity is not motivated by moral concern.

[10]Experimental economics shows that information about social history of past players has an influence on the behavior of current players (Berg *et al.*, 1995).

community all individuals positively condition their co-operation on an intangible stock of trust, and build it up, while in the other community everyone disregards the possible relevance of such a stock. Since there is neither physical capital accumulation nor changes in the parameters of the production and utility functions, the model is in a sense a repeated game, but trigger strategies are ruled out by assumption. A novel feature of the model is that without recourse to trigger strategies, there exists a continuum of contribution levels that can be sustained as steady states of feedback Nash equilibria. The set of these equilibria displays two important properties. First, the highest sustainable level of a flow public good is strictly between the static Nash level and the social optimum. Second, except at steady states, the level of voluntary contributions changes over time. Individuals gradually adjust their contributions, which converge to steady-state contribution levels. These properties are in sharp contrast to those obtained in the repeated game literature, where agents use threats. In that literature, different contribution levels can be supported by trigger strategies and the typical equilibrium is a constant level of contribution to the flow public good. The gradual building up of co-operation is a major contribution of the model of Benchekroun and Long (2008a).[11] In this model, agents are motivated only by self-interest. If individuals start with a very high level of co-operation, their co-operative behavior will decrease to a steady-state level. Similarly, beginning with a low level of co-operation, good behavior will increase to a steady-state level.

Let us provide more specific details of the model. There are two final goods: a private good, and a public good, the supply of which is denoted by G. (Here G is a flow, not a stock.) These goods are produced using labor. The community consists of n identical infinitely lived agents. There is no physical stock of productive input. This assumption allows a sharp focus on the role of social trust which is modelled as a stock, with its own law of motion. At time t, if agent i contributes $g_i(t)$ units of labor as input to the flow public good (such as cleaning a park), he can only produce $c_i(t) = c(g_i(t))$ units of the private good. The function $c(g_i)$ describes the player's production possibilities frontier. Assume $c'(g_i) < 0$, which implies that

---

[11]Other models capable of generating a dynamic path of co-operation are evolutionary games. In such models there must be at least two types of agents, with the proportion of the good type obeying an evolutionary process described by a replicator dynamics. In Carpenter (2004), an agent can either co-operate or free ride. In the long run, everyone becomes a free rider. This prediction has received some experimental support (Miller and Andreoni, 1991; Ledyard, 1995; Friedman, 1996; Carpenter and Matthews, 2005).

the quantity of private goods that a player can produce for consumption decreases when he increases his contribution of labor to the flow public good.

The instantaneous utility function of agent i is

$$u_i = U(G, c_i)$$

where G is the quantity of the public good. The utility function is increasing in both arguments. Let G_{-i} denote the total contribution of all other agents. Then, $G = G_{-i} + g_i$.

Agent i chooses a time path of contribution $g_i(t)$ to maximize his/her lifetime utility:

$$\int_0^\infty U(g_i(t) + G_{-i}(t), c(g_i(t)))e^{-rt}dt$$

If all agents believe that the level of voluntary contributions of other agents never changes, this lifetime optimization problem becomes in essence a static one, and the outcome is the static Nash equilibrium. The first-order condition of the static problem is

$$U_G(g_i + G_{-i}, c(g_i)) + U_x(g_i + G_{-i}, c(g_i))c'(g_i) = 0$$

This condition gives rise to the reaction function $g_i = R(G_{-i})$. A symmetric static Nash equilibrium is a vector $\mathbf{g}^N = (g^N, \ldots, g^N)$ such that

$$g^N = R(ng^N)$$

In this equilibrium, each individual's marginal rate of substitution is equated to his marginal rate of transformation:

$$MRS = \frac{U_G(ng^N, c(g^N))}{U_c(ng^N, c(g^N))} = -c'(g^N) = MRT \qquad (6.5)$$

There exists a unique symmetric static Nash equilibrium if the RHS of Eq. (6.5) is increasing in g, with $c'(0) = 0$ and the LHS is decreasing.[12] To find the social optimum one solves

$$\max_g U(ng, x(g))$$

---

[12]To ensure a diminishing marginal rate of substitution one may assume $U_{cG} \geq 0$, $U_{GG} \leq 0$ and $U_{cc} \leq 0$.

Let $\hat{g}$ denote the social optimal solution. The first-order condition for the social optimum is

$$nU_G(n\hat{g}, c(\hat{g})) + U_c(n\hat{g}, c(\hat{g}))c'(\hat{g}) = 0$$

Rearranging this equation, one gets the familiar Samuelsonian rule: the marginal rate of transformation must be equated to the sum of the marginal rates of substitution across all individuals.

$$\sum MRS_i = \frac{nU_G(n\hat{g}, c(\hat{g}))}{U_c(n\hat{g}, c(\hat{g}))} = -c'(\hat{g}) = MRT \qquad (6.6)$$

Benchekroun and Long (2008a) introduce a state variable $X(t)$ which is a summary of the social history, called the stock of co-operation, or trust. Assume that all individuals know the transition equation for the variable

$$\dot{X}(t) = G(t) - ng^N - \delta X(t) \qquad (6.7)$$

Here, $\delta > 0$ is the rate of depreciation of X. Equation (6.7) implies that if $X(0) = 0$ and $G(t) = ng^N$ always then $X(t)$ will remain at zero for ever, meaning the social capital will not be built up. On the other hand, as soon as $G(t)$ exceeds ng^N, the state variable will become positive.[13] If all agents always contribute to the flow public good at the socially desirable level $\hat{g}$ the resulting steady-state stock of co-operation will be $\widehat{X}_\infty = n\frac{\hat{g}-g^N}{\delta} > 0$.

A striking feature of the model is that the state variable enters neither the production function nor the utility function. One may thus say that this variable has no intrinsic value. Despite this lack of intrinsic value, $X(t)$ can indirectly influence the actual payoffs of all players. Suppose each agent i believes that all other players follow a rule of behavior $\mu_j(.)$ that conditions their contribution $g_j(t)$ to the stock of co-operation

$$g_j(t) = \mu_j(X(t)) \quad \text{for } j \neq i \qquad (6.8)$$

Assuming that agent i thinks that all other individuals have the same decision rule, we can drop the subscript j and write

$$G_{-i}(t) = (n-1)\mu(X(t))$$

Agent i has the following objective function

$$\max_{g_i} \int_0^\infty U(g_i(t) + (n-1)\mu(X(t)), c(g_i(t)))e^{-rt}dt$$

---

[13]If $X(0) = 0$ and agents never contribute less than the static Nash equilibrium level, $X(t)$ will never become negative.

subject to

$$\dot{X} = g_i + (n-1)\mu(X) - ng^N - \delta X, \quad X(0) = 0 \qquad (6.9)$$

In a symmetric equilibrium, all players act in the same way. What are the Markovian Nash equilibria for this differential game?

Benchekroun and Long (2008a) show that candidate equilibrium strategies must solve the differential equation

$$\mu'(X) = \frac{(r+\delta)\pi(\mu(X))}{(n-1)U_G + (n-1)\pi(\mu(X)) + \pi'(\mu(X))(n\mu(X) - ng^N - \delta X)}$$

$$(6.10)$$

where

$$\pi(\mu(X)) \equiv -U_G(n\mu(X), c(\mu(X))) - U_c(n\mu(X), c(\mu(X)))c'(\mu(X))$$

Analyzing the family of solutions of this differential equation, the authors prove that there exists a value $g^* \in (g^N, \hat{g})$ such that any level of contribution in the interval $[g^N, g^*)$ can be supported as a stable steady state by a profile of Nash equilibrium strategies.[14] This result is established without assuming that the utility function is quadratic.

Along any equilibrium path, individual contributions to the public good are changing over time. Agents gradually and smoothly adjust their contributions as the stock of cooperation X converges to its steady state level. In this model agents are informed about the characteristics of the game and know the strategies of other agents, yet the level of contribution, and the stock of co-operation, can only increase gradually.

The public good model of Benchekroun and Long (2008a) belongs to the class of models where the public good is "pure": agents do not care who contribute to the public good. We should note that many authors argue that some public goods have private components, known as the warm glow effect: the donor not only cares about the welfare level of the recipients, but also about the extent to which he has contributed to that welfare level. An obvious example of dynamic voluntary contributions to public goods with the warm glow effect is foreign aid: many developed countries non-cooperatively donate development aids to poor countries, and they seem to

---

[14]If $X(t)$ remains finite and U is bounded then the transversality condition is satisfied provided that

$$\lim_{t \to \infty} X(t) = \overline{X}$$

for some finite $\overline{X} > 0$.

care that part of the aid comes from them. For two dynamic game models of this non-pure public good type, see Kemp and Long (2009).

6.2. Contributions to a Discrete Public Good

Unlike the divisible public good case (such as a stock of library books), a discrete public good requires that the sum of contributions be exactly equal to a certain level. Kessing (2007) considers the case of a discrete public good, such as a bridge. Benefits start to flow only after the bridge has been completed. Further, it is assumed that to build the bridge the total effort required is a fixed number $\overline{S}$, no more, and no less. Under these conditions, what is the Nash equilibrium contribution of effort by each individual in a community consisting of n identical individuals?

This question can be posed in a static context, or in a dynamic context. In the static context, it has been found that there exists a continuum of asymmetric equilibria (Bagnoli and Lipman, 1989). In a dynamic context, this result can be replicated if agents play open-loop strategies. However, Kessing (2007) shows that with Markov-perfect strategies, the equilibrium is unique.

Kessing considers the following differential game. By normalization, before the bridge is built, each individual has a negative utility of $-B$ (where $B > 0$) per unit of time (e.g., because they have to swim across the river). After the bridge is built, their utility level is zero per unit of time.

Effort contributions involve a cost. If individual i contributes a flow of effort $E_i(t) \geq 0$ at time t, he incurs a cost $(1/2)E_i(t)^2$. Let $S(t)$ be the community's accumulated effort from time zero to time t:

$$S(t) = \int_0^t \sum_{i=1}^n E_i(\tau)d\tau$$

Then

$$\dot{S}(t) = \sum_{i=1}^n E_i(t), \quad S(0) = S_0$$

and the bridge is completed at time T where

$$S(T) = \overline{S} \quad \text{and} \quad S(0) = S_0 \text{ (given)}$$

where $\overline{S}$ is a fixed number. Assume $S_0 < \overline{S}$.

The welfare of individual i is then

$$V_i = \int_0^T e^{-rt}\left[-B - \frac{1}{2}E_i^2\right]dt + 0 \times e^{-rT}$$

6.2.1. *The co-operative solution*

Let us consider first the solution under a central planner. If the planner decides that no effort should be incurred, that is, $E_i(t)$ is chosen to be zero for ever, then the bridge is never built, and the welfare (or discounted stream of net benefits) of the community is simply $-nB/r \equiv \underline{W} < 0$. On the other hand, if the planner decides to build the bridge, the welfare of the community is the solution of the following constrained optimization problem

$$\max_{E_i,T} \int_0^T e^{-rt}\left[-nB - \frac{1}{2}nE_i^2\right]dt + 0 \times e^{-rT}$$

subject to the integral constraint[15]

$$\int_0^T nE_i(t)dt = \overline{S} - S_0 \tag{6.11}$$

Let λ be the (constant) shadow price associated with the integral constraint. The Hamiltonian is

$$H = e^{-rt}\left[-nB - \frac{1}{2}nE_i^2\right] + \lambda n E_i$$

The first-order condition gives

$$E_i(t) = \lambda e^{rt} \tag{6.12}$$

Since T is free, the transversality condition is

$$H(T) = e^{-rT}\left[-nB - \frac{1}{2}nE_i^2(T)\right] + \lambda n E_i(T) = 0$$

Replacing $E_i(T)$ by λe^{rT}, the transversality condition gives

$$\lambda = \frac{\sqrt{2B}}{e^{rT}} \tag{6.13}$$

---

[15] As shown in Leonard and Long (1992), for example, an integral constraint is equivalent to a differential equation with fixed boundary values. In our case, we have $\dot{S}(t) = nE_i(t)$ with $S(0) = S_0$ and $S(T) = \overline{S}$.

Next, we substitute Eq. (6.12) into the integral constraint to get

$$\lambda\left(e^{rT} - 1\right) = \frac{(\bar{S} - S_0)r}{n} \qquad (6.14)$$

Eliminating λ from Eqs. (6.13) and (6.14) we obtain

$$\frac{1}{e^{rT}} = \frac{n\sqrt{2B} - r(\bar{S} - S_0)}{n\sqrt{2D}} \qquad (6.15)$$

This equation has a finite solution T if and only if $n\sqrt{2B} - r(\bar{S} - S_0) > 0$, that is, the benefit per person, B, is high enough, or there are a large number of people in the community. If $n\sqrt{2B} - r(\bar{S} - S_0) < 0$, the optimal control problem (subject to the integral constraint (6.11)) would have no solution, and if $n\sqrt{2B} - r(\bar{S} - S_0) = 0$, T would be infinite, that is, the bridge would never be built.

Assume $n\sqrt{2B} - r(\bar{S} - S_0) > 0$. Then, the optimal T is

$$T = \frac{1}{r}\ln\left[\frac{n\sqrt{2B}}{n\sqrt{2B} - r(\bar{S} - S_0)}\right]$$

and the optimal shadow price is

$$\lambda = \frac{\sqrt{2B}}{e^{rT}} = \sqrt{2B} - \frac{(\bar{S} - S_0)r}{n}$$

The optimal initial effort is

$$E_i(0) = \lambda = \sqrt{2B} - \frac{(\bar{S} - S_0)r}{n} \qquad (6.16)$$

Since any time can be an initial time, Eq. (6.16) gives the optimal control in feedback form

$$E_i(z) = \sqrt{2B} - \frac{rz}{n} \qquad (6.17)$$

where z is the remaining necessary amount of cumulative effort, $z(t) \equiv \bar{S} - S(t)$, for all $S(t) < \bar{S}$.

Given that $n\sqrt{2B} - r(\bar{S} - S_0) > 0$, each individual's welfare obtained from implementing this optimal control problem is

$$V_i = -B\left(\frac{1 - e^{-rT}}{r}\right) - \frac{1}{2}\int_0^T e^{-rt}(\lambda^2 e^{2rT})dt$$

$$= -\frac{B}{r}\left(1 - \frac{1}{e^{2rT}}\right) > -\frac{B}{r} \equiv \underline{W}$$

Now let us compare the social optimal solution with the non-cooperative equilibria (in open-loop strategies, and in feedback strategies).

6.2.2. *Open-loop Nash equilibria*

It turns out that there is a continuum of open-loop Nash equilibria. At one extreme, there is an OLNE where the outcome is identical to the co-operative equilibrium (the social optimum) that we found above. At the other extreme, there is an OLNE where no individual contributes any effort, and hence the bridge is not built. There are also asymmetric equilibria, where identical agents do not behave identically. This result is similar to the result that, when there is a coupled constraint, it is likely that there is a continuum of equilibria (Rosen, 1965).[16]

Let us consider the optimization problem of individual i who takes as given the time path of the aggregate contribution of all other individuals $j \neq i$. Let us define

$$E_{-i}(t) = \sum_{j \neq i} E_j(t)$$

and

$$X_{-i}(t) = \int_0^t E_{-i}(t)dt$$

Given the time path of X_{-i}, individual i has the option of not contributing at all, or contributing until the total contribution equals $\overline{S}$ (at which time the bridge is completed).

Suppose that individual i wants to have the bridge built at some optimally chosen time T_i. His optimal control problem is then

$$\max_{E_i, T_i} \int_0^{T_i} e^{-rt} \left(-B - \frac{1}{2} E_i^2 \right) dt$$

subject to the integral constraint

$$\int_0^{T_i} E_i(t)dt = \overline{S} - S_0 - X_{-i}(T_i) \tag{6.18}$$

The Hamiltonian is

$$H_i = e^{-rt} \left(-B - \frac{1}{2} E_i \right) + \lambda_i E_i$$

---

[16]For a discussion of coupled constraints and Rosen's normalized equilibrium, see Chap. 2 on environmental economics.

Then, the necessary conditions are

$$E_i(t) = \lambda_i e^{rt}$$

$$H_i(T) = e^{-rT_i}\left(-B - \frac{1}{2}E_i^2(T)\right) + \lambda_i E_i(T) = 0$$

From these, we obtain

$$\lambda_i = \frac{\sqrt{2B}}{e^{rT_i}}$$

and

$$\lambda_i(e^{rT_i} - 1) = r(\overline{S} - S_0 - X_{-i}(T_i))$$

Clearly, if

$$X_{-i}(T_i) = \frac{(n-1)(\overline{S} - S_0)}{n}$$

then individual i's solution will be identical to what he does under the co-operative scenario of the preceding subsection. Then, his total contribution will be $(\overline{S} - S_0)/n$ and T_i will be identical to T found by the social planner. We have thus established the existence of an OLNE that coincides with the social optimum.

Let us turn to the other extreme and suppose that each individual i believes that all other individuals plan to contribute nothing, that is, $X_{-i}(t) = 0$ for all t. Then his optimization problem subject to the integral constraint (6.18) will have a solution if and only if $\sqrt{B} > r(\overline{S} - S_0)$. Assume $\sqrt{B} < r(\overline{S} - S_0) < n\sqrt{B}$. Then even though it would be socially optimal to have everyone contribute, there is an OLNE where nobody contributes.

The reader should verify that there are other OLNE, some of which involve asymmetric strategies: identical individuals do not choose the same strategy.

6.2.3. *Markov-perfect Nash equilibria*

Let us now show that there exists a MPNE in which individuals use the same strategy, and the outcome is socially inefficient: it takes a longer time to reach the required accumulated contribution level $\overline{S}$.

To find a MPNE, we must find for each player a strategy that conditions his contribution on the observed level of the accumulated stock. Clearly, since the marginal cost of contribution is small for contribution levels near zero, all players will have an incentive to contribute if the observed stock level S is almost equal to $\overline{S}$. It follows that a strategy that prescribes zero contribution for all stock levels cannot be individually rational. Taking this hint, we seek for each individual i a value function $V_i(S)$ that satisfies the HJB equation

$$rV_i(S) = \max_{E_i} \left\{ -B - \frac{E_i^2}{2} + V_i'(S)[E_i + (n-1)\phi_j(S)] \right\} \quad \text{for} \ \ S \in [0, \overline{S})$$

with the boundary condition $V_i'(\overline{S}) = 0$.

The maximization of the RHS of the HJB equation with respect to E_i yields the first-order condition

$$-E_i + V_i'(S) = 0$$

Considering a symmetric equilibrium, we can drop the subscripts i and j and set $\phi(E) = V'(S)$. Then

$$rV(S) = \left(n - \frac{1}{2} \right) (V')^2 - B$$

Define the variable y by

$$y = rV + B$$

so that

$$y = \frac{(y')^2}{b^2}$$

where

$$b \equiv \frac{r}{\sqrt{\left(n - \frac{1}{2} \right)}}$$

We have two differential equations

$$y' = b\sqrt{y}$$
$$y' = -b\sqrt{y}$$

Since we expect $V'(S) > 0$ for $S \in [0, \overline{S})$, we discard the second differential equation. Let $x = \sqrt{y}$, then the first differential equation becomes

$$\frac{dx}{dS} = \frac{1}{2} y^{-1/2} \frac{dy}{dS} = \frac{b}{2}$$

This gives $\sqrt{y(S)} = x(S) = (b/2)S + A$ where A is a constant of integration. The requirement that $V(\overline{S}) = 0$ implies $\sqrt{B} = \sqrt{y(\overline{S})} = (b/2)\overline{S} + A$. It follows that $A = \sqrt{B} - (b/2)\overline{S}$. Thus

$$\sqrt{y(S)} = -\frac{b}{2}(\overline{S} - S) + \sqrt{B} \quad \text{for } S \in [0, \overline{S}) \qquad (6.19)$$

For the RHS to be positive, we must assume that the following inequality holds

$$\sqrt{B} - (b/2)\overline{S} > 0 \qquad (6.20)$$

Thus, the following condition is necessary for the existence of a symmetric feedback Nash equilibrium leading to the bridge being built:

$$\sqrt{B} > \frac{r\overline{S}}{2\sqrt{\left(n - \frac{1}{2}\right)}} \qquad (6.21)$$

Let us compare this condition with its counterpart under the social planner (where we assume $S_0 = 0$ for simplicity):

$$n\sqrt{2B} - r\overline{S} > 0 \qquad (6.22)$$

Clearly whenever inequality (6.22) is satisfied, the condition (6.21) is also satisfied, but not vice versa.

Let us obtain some more information about the value function. From Eq. (6.19)

$$rV(S) = -\frac{b}{2}(\overline{S} - S)\left[2\sqrt{B} - \frac{b}{2}(\overline{S} - S)\right]$$

Thus, $V(S)$ is strictly convex. It reaches a minimum at

$$(\overline{S} - S) = \frac{2\sqrt{B}}{b}$$

that is, at some negative S. It follows that for all $S \in [0, \overline{S})$, $V'(S) > 0$.

The symmetric MPNE therefore consists of the strategy

$$E_i(S) = V'(S) = \frac{b}{2r}\left[2\sqrt{B} - b(\overline{S} - S)\right] = \frac{b\sqrt{B}}{r} - \frac{b^2}{2r}(\overline{S} - S) \qquad (6.23)$$

Substituting for b, we obtain the equilibrium feedback strategy

$$E_i(S) = \frac{2\sqrt{B\left(n - \frac{1}{2}\right)}}{2\left(n - \frac{1}{2}\right)} - \frac{r}{2\left(n - \frac{1}{2}\right)}(\overline{S} - S) \equiv \gamma S + \beta \qquad (6.24)$$

Let us compare this with the social planner's optimal feedback control rule

$$E_i^*(S) = \sqrt{2B} - \frac{r(\overline{S} - S)}{n}$$

Suppose that $S(0) = 0$. Under the feedback equilibrium, how long would it take for the stock of contribution to reach the level $\overline{S}$? Recall that $\dot{S} = nE_i = \gamma S + \beta$. Then $S(t) = -(\beta/\gamma) + Fe^{\gamma n t}$ where F is a constant that can be determined from $S(0) = 0$. So, $S(t) = -(\beta/\gamma) + (\beta/\gamma)e^{\gamma n t}$. Setting $S(T_m) = \overline{S}$ gives the time T_m at which the bridge is completed under MPNE

$$T_m = \frac{1}{\gamma n}\ln\left[1 + \frac{\gamma \overline{S}}{\beta}\right] = \frac{2 - (1/n)}{r}\ln\left[1 + \frac{r\overline{S}}{2\sqrt{B\left(n - \frac{1}{2}\right)} - r\overline{S}}\right]$$

Compare T_m with T under the social planner

$$T = \frac{1}{r}\ln\left[\frac{n\sqrt{2B}}{n\sqrt{2B} - r\overline{S}}\right] = \frac{1}{r}\ln\left[1 + \frac{r\overline{S}}{n\sqrt{2B} - r\overline{S}}\right]$$

Since $n\sqrt{2B} > 2\sqrt{B\left(n - \frac{1}{2}\right)}$ for $n > 1$, it follows that $T_m > T$. Thus, under the feedback Nash equilibrium, it takes more time to obtain sufficient fund to build the bridge.

Finally, one can prove the uniqueness of the feedback equilibrium, under the assumption that agents use only linear feedback strategies. See Kessing (2007) in this regard.

6.3. Corrective Taxation with Far-sighted Firms

Many economists would argue that one of the roles of the government is to ensure that market failures are corrected by fiscal or regulatory devices. The simplest textbook example of corrective taxation is the subsidization of a monopolist firm to induce it to produce output at the socially efficient level. Alternatively, if information about the cost structure is available, the regulator can simply require the firm to charge a price equal to the marginal cost and to satisfy demand at that price.[17] In the absence of the information about cost, alternative mechanisms can be developed based on information concerning the demand side.[18]

When only local information is available neither the marginal cost pricing coupled with a lump-sum transfer nor demand-based mechanism can be used. Benchekroun and Long (2008b) show that, when only local information is available, there exists a family of feedback subsidy rules that would (i) induce the monopolist to produce the optimal output *at all times*, and (ii) economize on the payments to the firm. They propose a subsidy rate that depends on a state variable which may be thought of as an index of the history of the monopolist's past performance. In Benchekroun and Long (2008b), the regulator is in effect a feedback Stackelberg leader while the monopolist is the follower in a dynamic game. The main contribution of Benchekroun and Long (2008b) is to build a continuum of dynamic incentive schemes that induce the agent to choose the socially optimal action. Each member of the family of dynamic incentive schemes results in a different distribution of surplus between the monopolist and the government. The multiplicity of efficiency-inducing schemes gives the regulator a high degree of freedom to pursue other objectives without compromising efficiency.

Before getting to the details, let us explain intuitively how the mechanism works. The regulator creates a "performance index" that summarizes the monopolist's past behavior. A high value of the index indicates a history of good performance. This index is continuously updated as the monopolist's output is observed. The subsidy rate paid to the monopolist at each moment is specified as an increasing function of the index. The formula for updating the index is announced from the outset, and thus the monopolist can optimally plan to build up the level of the index. This index may be regarded as an intangible asset that the firm can "invest" in. Technically, it is a state variable in the firm's optimal control problem. The index may also depreciate over time. The regulator decides on the depreciation rule to be applied to the index. The subsidy rate being dependent on the index, the monopolist knows that its output level at any

---

[17]A lump-sum subsidy may be necessary to satisfy the participation constraint.

[18]When the regulator has only information about the demand curve, Loeb and Magat (1979) propose that the monopoly be offered a subsidy equal to the consumer surplus at the price the monopoly charges. This approach has been refined by Sappington and Sibley (1989, 1990) and Vogelsang and Finsinger (1979) who propose incremental mechanisms to determine the amounts transferred to the monopolist. These mechanisms require global information about the demand curve.

given time t affects not only its current profit, but also its future profits through its influence on the future subsidy rates.

The authors show that there exists a continuum of such subsidy schemes. Each member of this continuum generates an infinite horizon dynamic optimization problem for the monopolist, and his profit-maximizing production path will be a constant path that yields at each instant the socially efficient output level.

Benchekroun and Long (2008b) begin their analysis with a monopolist in a static environment. Let x denote the firm's output. The total cost function is $C(x)$, and the inverse demand function is $p = p(x)$, with

$$C(0) = 0, C'(x) > 0, \qquad C'(0) < p(0), \quad p'(x) < 0$$

In the absence of a tax or subsidy, the profit function is

$$\Pi(x) \equiv p(x)x - C(x) \tag{6.25}$$

The function $\Pi(x)$ is assumed to be strictly concave and to attain its maximum at some $x^L > 0$. (The superscript L indicates that x^L is lower than the social optimum output.)

The social welfare is the area under the demand curve minus the production cost:

$$W(x) = \int_0^x p(y)dy - C(x)$$

Let $\hat{x}$ denote the socially optimal output level. At $\hat{x}$, marginal cost is equated to price

$$p(\hat{x}) = C'(\hat{x}) \tag{6.26}$$

It is assumed that $\hat{x}$ is unique.

The regulating agency only has local information about the cost and demand functions. Its main objective is to ensure that the output of the firm is exactly equal to $\hat{x}$. Assume that the regulator knows $\hat{x}$ and the value of the slope of the demand curve at $\hat{x}$. (Information about value of the slope of the demand curve at other output levels may not be available.)

The standard textbook mechanism to induce efficiency is to subsidize the monopolist's output at a *constant* rate s^* per unit of output, where

$$s^* = -p'(\hat{x})\hat{x} > 0 \tag{6.27}$$

Clearly s^* is the difference between the price $p(\hat{x})$ and the marginal revenue $p(\hat{x}) + p'(\hat{x})\hat{x}$ (both evaluated at $\hat{x}$). If the firm's output is at $\hat{x}$, its profit inclusive of subsidy is

$$\pi(\hat{x}) = \Pi(\hat{x}) + s^*\hat{x}$$

At $\hat{x}$, it holds that

$$\Pi'(\hat{x}) + s^* = 0$$

From a perpective that takes into account fairness, a major problem with the static subsidy rule is that the subsidy makes the monopolist even richer. Benchekroun and Long (2008b) propose a class of simple linear subsidy rules that do not enrich the monopolist as much and still ensure that he produces the socially optimal output level $\hat{x}$. Their proposal consists of devising an index Z that provides a measure of the cumulative performance of the monopolist. The initial value Z_0 is chosen by the regulator at the start of the game. The regulator announces a linear rule for updating $Z(t)$:

$$\dot{Z}(t) = x(t) - \delta Z(t) \qquad (6.28)$$

where $\delta > 0$ is the rate of depreciation of the index. The monopolist is informed of the differential equation (6.28), and of δ and Z_0. It is assumed that the regulator can commit that the rule will not be changed.

Benchekroun and Long (2008b) offer the following interpretation of (6.28). Suppose the regulator sets $Z(0) = \hat{x}/\delta$. Then when the firm's output is less than $\hat{x}$, the index $Z(t)$ will fall, indicating that the monopolist has "misbehaved", and that he will be punished by a future decrease in the subsidy rate. If the subsidy rule is well designed, the built-in threat of a decrease in the subsidy rate will induce the firm to choose $\hat{x}$ forever. Alternatively, the regulator may set $Z(0) < \hat{x}/\delta$. Then the firm knows that if it chooses $x = \hat{x}$, the index $Z(t)$ will rise. Facing the prospect of an increase in the subsidy rate, the monopolist will again choose the socially optimal output.

For concreteness, let us specify a subsidy rule of the form $S(x, Z) = \omega(X)x$, where $\omega(.)$ is a function defined for all $Z \geq 0$. This rule is linear in x and nonlinear in Z. At time t, the monopolist receives a subsidy $\omega(Z(t))$ per unit of output. Below is an example of the function ω. Let ρ be the interest rate. Define $\mu \equiv \rho/\delta + 1 > 1$. Let

$$\omega(Z) = s^* - JZ^{-\mu} \quad \text{and} \quad Z(0) = Z_0 > 0 \qquad (6.29)$$

Here, J is a positive number and s^* is given by Eq. (6.27). Benchekroun and Long (2008b) show that $\hat{x}$ will be achieved by appropriate choices of δ, J, and Z_0.[19] The family of rules specified by Eq. (6.29) ensures that if the monopolist builds up the level of the index, he will get a higher subsidy rate, since

$$\frac{d\omega}{dZ} = \beta J Z^{-\mu-1} > 0 \qquad (6.30)$$

Clearly, the subsidy rule (6.29) yields a negative subsidy rate (i.e., a tax) when the index $Z(t)$ falls below a certain threshold.

As the leader, the regulator informs the follower of the subsidy rule (6.29), the updating rule (6.28), the depreciation rate δ, the constant J, and the initial value Z_0. The firm's net profit is

$$\hat{\pi}(x, X) = p(x)x - C(x) + \omega(Z)x$$

The firm chooses a time path $x(t) \geq 0$ and a terminal time $T \geq 0$ to maximize the discounted stream of profit:

$$\max_{q \geq 0} \int_0^T \hat{\pi}(x, Z)e^{-\rho t} dt \qquad (6.31)$$

If the subsidy rule is not well designed, the firm could choose a *"hit and run"* strategy: to make a quick profit over some finite time interval $[0, T]$ and then exit the market. The firm will not hit and run if the subsidy rule and the parameters J, δ, and Z_0 are well chosen. Benchekroun and Long (2008b) give a numerical example to show that the monopolist will produce the socially optimal output, while receiving much lower subsidies than under the conventional static mechanism. They also show that there is a continuum of subsidy rules that perform the same efficiency-inducing role.

In the above model, there is no physical state variable, only an intangible asset, the index Z. It turns out that the multiplicity of efficiency-inducing tax/subsidy schemes also arises when there is a natural state variable (e.g., the stock of pollution) that is affected by the firm's output (Benchekroun and Long, 2002b). In such cases, an artificial index Z is not necessary. The tax/subsidy rule can be conditioned on the physical stock, for example, the stock of pollution. This result for monopoly can be

---

[19] J has to be within a certain range to ensure that the firm is willing to participate, while δ and Z_0 must be chosen to ensure that the monopolist does not exit the market in finite time.

extended to the case of oligopoly (Benchekroun and Long, 1998).[20] Related taxation issues in the context of monopoly extraction of an exhaustible resource are discussed by Bergstrom *et al.* (1982) using a model of open-loop Stackelberg equilibrium (OLSE) (where the regulator is a leader who announces a time path of tax rate), and by Karp and Livernois (1992) using feedback tax rule. The problem with the open-loop tax proposed by Bergstrom *et al.* (1982) is that a sophisticated resource-extracting monopolist would be able to manipulate it to his advantage. The mechanism proposed by Karp and Livernois (1992) for a resource-extracting monopolist overcomes the manipulation incentives.

   Another stream of literature is concerned with dynamic Pigouvian tax in the context of global warming, see Wirl (1994), Rubio and Escriche (2001), Liski and Tahvonen (2004), Fujiwara and Long (2009c). We refer the reader to Chap. 2 for a fuller discussion of these models.

6.4. Redistributive Taxation

Judd (1985) and Chamley (1986) model redistributive taxation as a differential game between an infinitely lived government (the leader) and a continuum of identical infinitely lived individuals (the followers). They use the concept of OLSE. One of their results is that in the long run, the optimal tax on capital income is zero. Kemp *et al.* (1993) point out that Judd's open-loop solution may not converge to a steady state. Using bifurcation analysis, they show that optimal open-loop taxation may be cyclical.[21]

   As we have pointed out in Chap. 1, OLSE are generically time-inconsistent. Policy recommendations based on analysis of OLSE should therefore be interpreted with care. Another point concerning the analysis of Chamley and Judd is that they assume that the government can influence the representative capitalist's initial shadow price of capital. As Xie (1997) points out, that assumption cannot be satisfied under certain specification of utility and production functions. He provides an example where the representative capitalist's shadow price of capital turns out to be independent of the time path of taxes. In that example, the elasticity of output with respect to capital just happens to be equal to the elasticity of marginal utility with respect to consumption. Lansing (1999) and Long and Shimomura (2000, 2001) examine the case where capitalists have a

---

[20]In their 1998 paper, there was no mention of this multiplicity.
[21]They also consider a feedback Stackelberg solution, assuming that individual capitalists are too small to recognize that their savings will influence the aggregate capital stock.

logarithmic utility function and their income is linear in their asset holding. In this case, the elasticity of income with respect to the asset is unity, and so is the elasticity of marginal utility. They find that the Chamley–Judd assumption fails to hold under these conditions, and prove that in this case the optimal tax on capital income is nonzero, even in the long run. Mino (2001b) and Karp and Lee (2003) contribute a further result: they prove that the government's open-loop policy is time-consistent if and only if the value function of the followers (say the capitalists) is additively separable in the state variable and in time.

Mino (2001a) extends the Chamley–Judd model to economies with increasing returns. He shows that if there exists a stable OLSE (where the government is the leader) the optimal tax on capital income is negative. However, he points out that this negative tax cannot arise in the feedback Stackelberg case if a balanced budget is required at each point of time. On the other hand, if the government can borrow and lend then a negative capital income tax rate in the steady state can arise in a feedback Stackelberg equilibrium.

6.5. Altruism, Distributive Justice, and Intergenerational Equity

Intergenerational equity issues have given rise to much academic debate among economists as well as philosophers and other social scientists. Environmentalists argue that the welfare of future generations is seriously compromised by the current rate of exploitation of natural assets, and that the market mechanism fails to take into account moral concerns about unfairness to our descendents. At the other pole, however, some philosophers have pondered about the apparent disparity of standards of living across generations in human history. In particular, Kant (1784) found it disconcerting that earlier generations should carry the burdens for the benefits of later generations,[22] and Rawls (1999, p. 253) was concerned that

---

[22]In his 1784 essay, "Idea for a Universal History with a Cosmopolitan Purpose", Kant put forward the view that nature is concerned with seeing that man should work his way onward to make himself worthy of life and well-being. He added: "What remains disconcerting about all this is firstly, that the earlier generations seem to perform their laborious tasks only for the sake of the later ones, so as to prepare for them a further stage from which they can raise still higher the structure intended by nature; and secondly, that only the later generations will in fact have the good fortune to inhabit the building on which a whole series of their forefathers . . . had worked without being able to share in the happiness they were preparing". See Reiss (1970, p. 44).

"the utilitarian doctrine may direct us to demand heavy sacrifices of the poorer generations for the sake of greater advantages for later ones that are far better off". He argued that even if one cannot define a precise just savings principle, one should be able to avoid these sorts of extremes.

Some economists believe that intergenerational equity can be served by obeying the maximin principle, which aims at maximizing the welfare level of the worse off generation from the present time to the indefinite future. In this context, I would like to point out that while the maximin principle of distributive justice has been attributed to Rawls, many economists did not realize that Rawls advocated this principle of justice only in a static setting. He emphatically stressed that such a principle should not be applied to problems of intergenerational equity.[23]

As is well known, the maximin principle leads to zero saving. Without capital accumulation, there is little chance that just institutions can be developed and sustained. Rawls (1971, 1999) argues that to obtain reasonable results, the maximin principle must be modified: it must be supplemented by a just savings principle. He acknowledges that it is difficult to formulate a just savings principle. At the same time, certain reasonable assumptions would set limits on the savings rate. Thus, in dealing with intergenerational equity, Rawls proposes that the hypothetical parties in the original situation of "veil of ignorance" are heads of family, and that the principle adopted must be such that the parties wish all earlier generations to have followed it. "Thus imagining themselves to be fathers, say, they are to ascertain how much they should set aside for their sons and grandsons by noting what they would believe themselves entitled to claim of their fathers and grandfathers".[24] The heads of family behind the veil of ignorance are to be "regarded as family lines, with ties of sentiment between successive generations" (Rawls, 1971, p. 292).

Parental altruism, mentioned by Rawls, can lead to a dynamic game among generations. Suppose a father cares about the consumption level of his son, but not about that of his grand son. How much should he save for his son? This decision cannot be made without knowing how his son is going to allocate his lifetime resources between consumption and bequest. Thus a father's consumption-saving strategy depends on his son's strategy, which in turn would depend on the strategy of the latter's offsprings, and so on.

---

[23]Long (2008) and Alvarez-Cuadrado and Long (2009) argue that many economists have misunderstood Rawls. They point out that Rawls was well aware that maximin would imply no savings at all, and was strongly opposed to the application of the maximin principle to intergenerational distributive justice.

[24]Rawls (1999) p. 256.

The time path of realized savings of a family is thus a Nash equilibrium of a dynamic game involving an infinite sequence of decision makers.

Games among generations have been a lively area of research. This literature springs from the so-called Arrow–Dasgupta model (Arrow, 1973, Dasgupta, 1974b). Noting the Rawlsian concern that the unmodified maximin principle would imply zero saving, Arrow and Dasgupta propose a model where each individual lives for just one period, and each generation's welfare is an additively separable function of its own consumption and the consumption of the next generation:

$$W_t(C_t, C_{t+1}) = U(C_t) + \beta U(C_{t+1})$$

where $\beta \leq 1$ and $U(.)$ is strictly concave. Asheim (1988) calls this formulation "paternalistic altruism".[25] Arrow and Dasgupta assume that the technology is linear

$$K_{t+1} = \alpha(K_t - C_t) \quad \text{where} \quad \alpha > 1$$

The objective function of the social planner at time $t = 0$ in the Arrow–Dasgupta model is to find a feasible infinite sequence $(C_0, C_1, \ldots, C_t, \ldots)$ and the greatest possible number $\overline{W}$ such that the welfare of every generation from $t = 0$ onward will be at least as high as $\overline{W}$:

$$W_t(C_t, C_{t+1}) \geq \overline{W} \quad \text{for all } t \in \{0, 1, 2, \ldots\}$$

Such a solution $\overline{W}$ is called the maximin level of welfare. This formulation allows growth to take place.[26] As pointed out by several authors (Dasgupta, 1974b; Calvo, 1978; Leininger, 1985), the Arrow–Dasgupta model has the property that the solution of the optimization problem is generally time-inconsistent: if the planner can replan at some time $t_1 > 0$, he/she would not choose the same sequence. This is because the maximin principle is, in the words of Asheim (1988), "ancestor-insensitive": for the social planner who plans in period $t + 1$, the welfare levels of the individuals who lived in periods t, $t - 1$, $t - 2$ etc. have no relevance. This property is the main source of the time-inconsistency, which arises even in more general models, where W_t is nondecreasing in each C_{t+j} where $j = 1, 2, \ldots, n$.

---

[25]The adjective "paternalistic" here indicates that the father cares about his son's consumption level, instead of his son's own view of his welfare which may include the grandson's utility etc.

[26]Arrow (1973, Theorem 2) showed that the maximin program can display a saw-tooth shaped path of consumption and utility. This was generalized by Suga (2004).

One way of avoiding time-inconsistency (at least in the context of the familiar one-sector neoclassical technology) is to assume non-paternalistic altruism: under this form of altruism, the welfare of an individual is the sum of a function of his current consumption and the welfare (rather than utility of consumption) of the next generation, as viewed by the latter. Calvo (1978) and Rodriguez (1981) use a recursive form of what Asheim (1988) calls "non-paternalistic altruism". Denote by $_tC$ the sequence of consumption levels of succesive generations starting from period t:

$$_tC = (C_t, C_{t+1}, C_{t+2}, C_{t+3}, \ldots)$$

and similarly for other sequences. Under non-paternalistic altruism, the preferences of the representative member of generation t are represented by

$$U(_tC) = v(C_t) + \beta U(_{t+1}C) = \sum_{s=t}^{\infty} \beta^{s-t} v(C_s)$$

where $0 < \beta < 1$. Asheim (1988) calls $v(.)$ the stationary one-period felicity function, and $U(.)$ an ancestor-insensitive stationary altruistic utility function. The parameter β is the felicity discount factor. Under this specification of utility, Calvo and Rodriguez find that in the Arrow–Dasgupta one-asset model the maximin program is time-consistent.

However, Asheim (1988) shows that if technology includes both a man-made capital stock and a stock of non-renewable resource, then time-inconsistency reappears. This motivates setting up a game-theoretic model. In such a model (see, e.g., Lane and Mitra, 1981; Leininger, 1986; Asheim, 1988), the current generation chooses a strategy which is a best reply to the strategies of future generations.

Asheim's main result is that in this game-theoretic setting, the equilibrium programs have the property of maximizing altruistic utility over the class of feasible programs with non-decreasing consumption. In Asheim's paper, generations retain their commitment to maximin as an ethical principle. In other words, Asheim assumes that the payoffs (as opposed to utility) of generation t is an ancestor insensitive and stationary social welfare function defined as the utility level of the least advantaged generation:

$$W(_tC) = \inf_{s \geq t} U(_sC)$$

This is in sharp contrast to the usual game-theoretic models where authors assume that each generation maximizes its altruistic utility.[27]

Let Y_t denote the total output (inclusive of the non-depreciated opening stock of man-made capital) and X_t denote the stock of non-renewable resource. Given (Y_t, X_t), generation t seeks to maximize $W(_t C)$. Asheim shows that the maximin program at time zero is time-consistent if and only if the discount factor β is small enough. Thus, myopia (in the form of a high discount rate, i.e., low discount factor β) helps time-consistency. Time-inconsistency arises if and only if each generation prefers the next generation to consume on its behalf.

In the case of time-inconsistency, it is natural to turn to the concept of Markov-perfect equilibrium. Then each generation behaves in a sophisticated way, choosing the best program among the class of programs that it knows future generations will carry out. A generation's equilibrium strategy is a best reply to the equilibrium strategies of succeeding generations. Assuming that as soon as a time-consistent program exists, it will be immediately followed, Asheim (1988) proves that there exists a unique subgame-perfect equilibrium, where equilibrium strategies are stationary and Markovian (in the sense that each generation's action is uniquely determined by the inherited levels of state variables).[28] Interestingly, the equilibrium program maximizes the altruistic utility over the set of feasible programs with non-decreasing consumption.

Dasgupta (1974b) explores the implications of some alternative criteria for justice between generations. Some more recent alternative criteria are Chichilnisky's sustainable preference (Chichilnisky, 1996) which includes the axioms of nondictatorship of the present, and nondictatorship of the future, and the Mixed Bentham–Rawls criterion proposed by Alvarez-Cuadrado and Long (2009).

For a dynamic game between two nations that adopt intergenerational equity objectives, see Long (2006b).

---

[27]See Dasgupta (1974b), Kohlberg (1976), Lane and Leininger (1984), Lane and Mitra (1981), Leininger (1986), Bernheim and Ray (1983, 1987), and Pezzey (1994).

[28]Instead of assuming that "as soon as a time-consistent program exists, it will be immediately followed", it might be more desirable to prove that this is a property of the equilibrium, under some additional game-theoretic requirements. One such approach is suggested by Asheim, using the concept of standard of behavior (Greenberg, 1986), and imposing internal and external stability in the sense of a von Neumann–Morgenstern abstract stable set.

6.6. Fiscal Competition and Electoral Incentives

6.6.1. *Factor mobility and fiscal competition*

In their analysis of redistributive taxation, Chamley and Judd restrict their attention to a closed economy. In practice, a tax imposed on the earnings of a factor of production will encourage that factor to move to another jurisdiction where the tax rate is lower.[29] Stigler (1965) points out that if all factors are mobile, redistributive taxation in a multijurisdiction world is practically infeasible.

A number of two-period models have been developed to investigate the implications of simple dynamic games between the owners of partially mobile factors of production on the one hand, and a local government that tries to redistribute income in favor of some group, on the other. Lee (1997) shows that if capital movements involve adjustment costs, there will be a wedge between the internal rate of return and the external one. Jensen and Thomas (1991) model a game between two governments that use debt policies to influence the intertemporal structure of taxation. Huizinga and Nielson (1997) formulate a two-period model in which even though capital is perfectly mobile, foreign capitalists in effect earn rents from local immobile resources.

Wildasin (2003) presents a continuous-time, infinite-horizon model in which infinitely lived agents react to changes in taxation by moving resources across jurisdictions. Adjustment costs are explicitly taken into account. The author focuses on the case of a once-over tax change, and does not deal with optimal time-varying tax rates. Instead, the analysis emphasizes the costly adjustment process, and draws on the adjustment cost literature in macroeconomics (Turnovsky, 2000). The main point is that capital earns quasi-rent which can be taxed away, but such quasi-rents erode with time. The optimal capital income tax rate depends crucially on the degree of capital mobility. Wildasin does not model a dynamic game involving the competition among jurisdictions to attract mobile resources.

Köthenbürger and Lockwood (2007) model a dynamic game of tax competition among n regional governments. The government of region i uses a source-based capital income tax τ_i to finance its provision of a local public good. Each government chooses the time path of τ_i to maximize the lifetime welfare of its representative infinitely lived resident. The authors assume that the output of each region is subject to stochastic shocks, represented by Brownian motions. This uncertainty ensures that capital does not stampede

---

[29]With the exception of completely immobile factors, such as land and mineral resources.

out of a region when its government increases the capital income tax rate.[30] Because of the desire for portfolio diversification, each household spreads its wealth among all regions. In this model, when governments compete by setting tax rates, on top of the usual fiscal externality associated with capital mobility, there is a "risk exposure externality": tax rates impact the investor's exposure to risks.

Due to the assumptions of logarithmic utility and linear technology, the value function of each government is dependent only on the stock of wealth of its residents. Under uncertainty, the equilibrium Markov-perfect tax strategy is a positive constant (while it would be zero under certainty due to the race to the bottom). The authors find that in the presence of uncertainty the growth rate under centralized tax setting is higher than under decentralized tax setting.

6.6.2. *Electoral incentives, corruption, and local public goods*

Another stream of literature on the competition among jurisdictions focuses on the relationship between rent-seeking and electoral incentives. Consider, for example, a country with a federal system of government. In deciding whether to reelect the incumbent party of a state or provincial government, voters may compare the performance of their government with that of an adjacent state. The outcome of course depends on both the electorate's information about the incumbent and their perception of the nature of governments.

There are nowadays very few adherents to the traditional view of governments as benevolent maximizers of social welfare. On the other hand, there is no dominant alternative theory. At one extreme, there is the conception of the government as a Leviathan, maximizing its size with extreme gluttony. At the other extreme, there is the Chicago School's view that political competition aligns politicians' interest with that of the majority of voters (otherwise, they would lose elections).

Some economists use the agency model, casting voters as principals and politicians as agents. Under such a model, rent appropriation by politicians can occur under asymmetric information. Barro (1973) and Ferejohn (1986) present models of political behavior under moral hazards, under the assumption that voters cannot perfectly observe politicians'

---

[30]Leour and Verbon (1977) consider a similar model under certainty. They assume that investors want to diversify their portfolios.

efforts. This informational asymmetry allows politicians to capture some rent. In a dynamic context, it is optimal for voters to offer some rent to reelection-seeking politicians to avoid higher rent extraction in the current period. Others (Rogoff and Sibert, 1988; Rogoff, 1990) emphasize the role of adverse selection and model the associated signalling game. Before the elections, politicians try to signal their competence, and engage in lower rent extraction than in the post-election period. According to this view, the role of elections is to choose the right type of politicians.

How does the electorate evaluate the incumbent politicians? A recent strand of literature, known as theory of yardstick competition, has received theoretical and empirical support. The basic idea is as follows: voters judge their incumbent goverment using as a benchmark the performance of another government in a comparable state. Inferring from data on the US state elections, Besley and Case (1995) find that the theory of yardstick competition can explain actual vote-seeking and tax-setting patterns. However, Besley and Smart (2001) show that the effect of yardstick competition on voters' welfare is ambiguous. While such competition allows the punishment of bad incumbents, it also leads to more rent extraction by incumbents who anticipate they will lose. Belleflame and Hindricks (2005) show that yardstick competition cannot fail if the reelection effect is strong. Ellis *et al.* (2005) find evidence that yardstick competition among politicians reduces corruption.

Long and Sengupta (2008) present a dynamic model where each incumbent government faces an election after a fixed term. The stock of reputation can build up or deteriorate over time. The incumbent government cares for both instantaneous and end-of-term reputation. The authors show that under unitary performance evaluation, dynamic incentives restrain the politician only if the shadow value of reputation (which measures current and future marginal benefits of increased reputation) is sufficiently high throughout the term. For such a high shadow value to exist, benefits of both instantaneous and end-of-period reputations have to be high enough. On the other hand, under relative performance evaluation, two adjacent governments are playing a dynamic game. In comparison to the static case, dynamic incentives impose more restriction on rent appropriation. When relative evaluations matter, a higher shadow price of reputation in one region reduces the incentive of rent appropriation in the *other* region as well.

The model is simple. Let τ_i be the tax collected from province i, and r_i be the rent appropriated by the government of that province. The amount

$(\tau_i - r_i) \equiv g_i$ is used to supply a local public good. The utility function of the voters in province i is

$$U(g_i, \tau_i) = u(g_i) - D(\tau_i)$$

It is assumed that $u(.)$ is concave and increasing, and $D(.)$ is convex and increasing. The function $D(.)$ includes the cost of forgone consumption of the private goods and the distortionary costs of taxation.

The government's private benefit from rents is denoted by $B(r_i)$. The stock of reputation is represented by a state variable, R_i. The evolution of this stock follows a simple law:

$$\dot{R}_i(t) = u(g_i(t)) - D(\tau_i(t)) - \gamma_i \Omega_i(r_i(t), r_j(t)), \quad \gamma_i > 0$$

The parameter γ_i represents the degree of voters' sensitivity to relative performance. The function $\Omega_i(r_i(t), r_j(t))$ indicates that relative performance is measured by comparing rents. Every T years, there is an election. The incumbent seeks to maximize the sum of the government's benefits over the current term of office and its expected post-election benefits:

$$\int_0^T e^{-\delta_i t}[B(r_i(t)) + \omega_i R_i(t)]dt + e^{-\delta_i T}V(R_i(T))$$

Here $V(R_i(T))$ is the government's expected value of post-election gains and ω_i refers to the weight politicians assign to their stock of reputation. The chance that the incumbent is reelected depends on the stock of esteem the electorate has for him at the election time.

To understand the importance of yardstick competition, let us consider first the case where it is absent, such that r_j does not appear in the function Ω_i. Assume that $\Omega_i(r_i)$ is convex and $V(R_i)$ is linear, that is, $V(R_i(T)) = p_i R_i(T)$ where $p_i > 0$.

Let ψ_i be the co-state variable associated with the state variable R_i. The Hamiltonian for the incumbent government is

$$H = B(r_i) + \omega_i R_i + \psi_i[u(\tau_i - r_i) - D(\tau_i) - \gamma_i \Omega(r_i)]$$

One obtains the necessary conditions

$$\psi_i[u'(\tau_i - r_i) - D'(\tau_i)] = 0 \tag{6.32}$$

$$B'(r_i) - \psi_i[u'(\tau_i - r_i) +_i v\Omega'(r_i)] = 0 \tag{6.33}$$

$$\frac{d\psi_i}{dt} = \delta_i \psi_i - \omega_i \tag{6.34}$$

In particular, the transversality condition is

$$\psi_i(T) = p_i \qquad (6.35)$$

This equation states that the shadow price of R_i at the election time is equated to the marginal contribution of reputation to the post-election value function. The differential equation (6.34) and the transversality condition (6.35) yield the optimal time path of $\psi_i(t)$:

$$\psi_i(t) = \frac{\omega_i}{\delta_i} + K_i e^{\delta_i t} \qquad (6.36)$$

where K_i is the constant of integration, determined by using the transversality condition (6.35):

$$K_i = \left(p_i - \frac{\omega_i}{\delta_i} \right) e^{-\delta_i T}$$

Therefore, the optimal time path of the shadow price is

$$\psi_i^*(t) = \frac{\omega_i}{\delta_i} [1 - e^{-\delta_i (T - t_i)}] + p_i e^{-\delta_i (T - t_i)} > 0 \qquad (6.37)$$

Its rate of change can be positive or negative:

$$\frac{d\psi_{i*}}{dt} = \delta_i \left(p_i - \frac{\omega_i}{\delta_i} \right) e^{-\delta_i (T - t_i)}$$

Let us turn to the time path of rent $r_i(t)$. Note that Eq. (6.32) shows that the tax $\tau_i(t)$ can be expressed as a function of $r_i(t)$, independently of the shadow price

$$\tau_i(t) = \tau(r_i(t)) \qquad (6.38)$$

In particular, its derivative is bounded between zero and one:

$$0 < \tau_i'(r_i) = \frac{u''}{u'' - D''} < 1 \qquad (6.39)$$

Equations (6.38) and (6.33) show that the chosen level of rent is a function of the optimal shadow price ψ_i^*:

$$\frac{B'(r_i)}{u'(\tau_i(r_i) - r_i) + \gamma_i \Omega'(r_i)} = \psi_i^* \qquad (6.40)$$

From Eq. (6.40), the rate of change of the rent is

$$\frac{dr_i}{dt} = \frac{u' + \gamma_i \Omega'}{\{B'' - \psi_i^*[\gamma_i \Omega'' + u''(1 - \tau_i')]\}} \frac{d\psi_i^*}{dt} \qquad (6.41)$$

The term inside the curly brackets $\{\cdots\}$ is negative. It follows that the rent $r_i(t)$ is increasing over time if and only if the optimal shadow price is decreasing over time, that is, if and only if $\delta_i p_i < \omega_i$.

Comparing Eq. (6.40) with what would be obtained in the static case, one finds that the level of rent in the dynamic model is lower than that obtained in the corresponding static model if and only if the shadow price is greater than 1. This will be the case if $p_i > 1$ and $\psi^*(t)$ is falling over time, that is, $\delta_i p_i < \omega_i$. It follows that in the case of a unitary system of government, if the incumbent politician is patient and attaches a high terminal value p_i per unit of reputation and a high weight to instantaneous esteem then the dynamic reelection incentives restrict the politician's rent-seeking relative to the static case.

Turning to the analysis of the effect of relative evaluation, Long and Sengupta (2008) specify the law of motion

$$\frac{dR_i}{dt} = u(\tau_i - r_i) - D(\tau_i) - \gamma_i \Omega_i (r_i - \kappa_i r_j)$$

Here, $\kappa \geq 0$ is a parameter. If κ is strictly positive, the rate of change in the stock of esteem R_i depends not only on r_i but also on r_j. Therefore, with $\kappa > 0$, the optimal time path of rent extraction by one provincial government depends on what it expects the rent extraction path of the neighboring government will be. Thus, the two provincial governments are engaged in a differential game. Since the game is linear in the state variables, the OLNE and the feedback Nash equilibrium are the same.[31] The optimality conditions for government i are

$$u'(\tau_i - r_i) - D'(\tau_i) = 0 \qquad (6.42)$$

$$B'(r_i) = [u'(\tau_i - r_i) + \gamma_i \Omega_i'(r_i - \kappa_i r_j)]\psi_i \qquad (6.43)$$

$$\frac{d\psi_i}{dt} = \delta_i \psi_i - \omega_i \qquad (6.44)$$

$$\psi_i(T) = p_i \qquad (6.45)$$

The Nash equilibrium time paths of the shadow prices are then

$$\psi_i^*(t) = \frac{\omega_i}{\delta_i} + \left(p_i - \frac{\omega_i}{\delta_i}\right) e^{-\delta_i(T-t)} \qquad (6.46)$$

---

[31]See Docker *et al.* (2000) for a proof of this result.

Hence

$$\frac{B'(r_i)}{u'(\tau_i(r_i) - r_i) + \gamma_i \Omega_i'(r_i - \kappa_i r_j)} = \psi_i^* \tag{6.47}$$

This equation yields the reaction function

$$r_i = \widehat{r}_i(r_j, \psi_i^*, \kappa_i) \tag{6.48}$$

The derivative $d\widehat{r}_i/dr_j$ can be obtained from

$$\{B''(r_i) - \psi_i^* u''[\tau_i' - 1] - \psi_i^* \gamma_i \Omega_i''\} dr_i + \kappa_i \psi_i^* \gamma_i \Omega_i'' dr_j = 0$$

If $\kappa_i > 0$, then the slope of the reaction function is positive and less than unity:

$$1 > \frac{d\widehat{r}_i}{dr_j} = \frac{\kappa_i \psi_i^* \gamma_i \Omega_i''}{\psi_i^* \gamma_i \Omega_i'' + [\psi_i^*(\tau_i' - 1)u'' - B_i'']} > 0$$

This equation shows that the rents are strategic complements. Note that an increase in the shadow price will shift the reaction curve downward:

$$\frac{\partial \widehat{r}_i}{\partial \psi_i^*} = -\frac{\gamma_i \Omega_i'}{\psi_i^* \gamma_i \Omega_i'' + [\psi_i^*(\tau_i' - 1)u'' - B_i'']} < 0$$

The intersection of the two reaction curves $r_i = \widehat{r}_i(r_j; \psi_i^*, \kappa_i)$ and $r_i = \widehat{r}_j(r_i, \psi_j^*, \kappa_j)$ gives the rents obtained at time t, where $\psi_i^*(t)$ and $\psi_j^*(t)$ are the Nash equilibrium shadow prices.

How does an increase in p_j (which increases $\psi_j^*(t)$) affect the equilibrium rents at time t in both provinces? Using Eq. (6.48), consider the system of equations

$$r_1^N(t) - \widehat{r}_1(r_2^N(t), \psi_1^*(t), \kappa_1) = 0$$
$$r_2^N(t) - \widehat{r}_2(r_1^N(t), \psi_2^*(t), \kappa_2) = 0$$

Here, the superscript N indicates the Nash equilibrium value.

Holding $\psi_1^*(t)$ constant, differentiate the system with respect to $\psi_2^*(t)$

$$\begin{bmatrix} 1 & -\dfrac{\partial \widehat{r}_1}{\partial r_2^N} \\ -\dfrac{\partial \widehat{r}_2}{\partial r_1^N} & 1 \end{bmatrix} \begin{bmatrix} \dfrac{dr_1^N}{d\psi_2^*} \\ \dfrac{dr_2^N}{d\psi_2^*} \end{bmatrix} = \begin{bmatrix} 0 \\ \dfrac{\partial \widehat{r}_2}{\partial \psi_2^*} \end{bmatrix}$$

Hence

$$\frac{dr_1^N}{d\psi_2^*} = \frac{1}{\Delta}\frac{\partial \widehat{r}_2}{\partial \psi_2^*} < 0$$

where

$$\Delta \equiv 1 - \frac{\partial \widehat{r}_2}{\partial r_1^N}\frac{\partial \widehat{r}_1}{\partial r_2^N} > 0$$

It follows that the higher the neighboring incumbent's shadow value of esteem, the lower the $r_i(t)$ in province i.

The following testable hypotheses emerge from this dynamic game model. First, in a democracy where elections are *very* frequent, incumbents may be more corrupt than in a country where the incumbents are appointed for a longer term. Second, among countries where the incumbents are appointed for a longer term, politicians in a federal country (where the voters employ relative performance evaluation) are likely to be less corrupt than their unitary-nation counterparts.

Finally, note that in the model of Long and Sengupta (2008), the public good is a flow. An obvious entension is to consider stock public goods. If politicians' reputation depends on both public good and rent, a "stock" public good may allow for more rent diversion.

Chapter 7

DYNAMIC GAMES IN MACROECONOMICS

This chapter surveys applications of the theory of dynamic games to macroeconomic policies, growth, and distribution. An earlier survey of this literature was provided by Pohjola (1986). Our main emphasis will be on the more recent literature. In Section 7.1 we focus on macroeconomic policies under rational expectations and international macroeconomic policies. Section 7.2 turns to the questions of wealth distribution, the struggles that determine the endogenous evolution of property rights regime, and their effects on growth rates. Section 7.3 deals with the conflicts between capitalists and workers, and their effects on economic growth.

7.1. Macroeconomic Policies

7.1.1. *Dynamic games between a policymaker and the private sector*

The papers by Kydland and Prescott (1977) and Calvo (1978) brought to macroeconomists something that microeconomists had known before: when non-dominant players have rational expectations, dynamic optimization by a dominant player using open-loop control leads to time-inconsistency. The recognition of this type of time-inconsistency appeared in a number of microeconomic models, for example, Coase (1972), Sieper and Swan (1973), as well as in the game-theoretic papers of Simaan and Cruz (1973b) and Aumann (1973).

As we have explained in Chap. 1, the time-inconsistency of open-loop Stackelberg equilibria (OLSE) is reflected in the fact that, under certain regularity conditions (Xie, 1997; Dockner *et al.*, 2000), the Stackelberg leader's optimal shadow price of the follower's co-state variable is zero at the beginning of the planning horizon and is typically non-zero afterward. At a later stage, if the leader (for example, a central bank) is released from its initially chosen path, it will want that shadow price to be zero again, implying a change in the path of its control variable (Driffill, 1982; Buiter, 1983). Time-inconsistent policies fail to satisfy Bellman's principle of optimality, because one of the state variables of the open-loop Stackelberg leader (namely, the private sector's co-state variable associated with its capital stock) is a jump state variable. For example, in the context of fiscal policies, this co-state variable is dependent on the private sector's expectation of the future path of tax rates.

It was thought that a major implication of this time-inconsistency issue is that optimal control theory should not be used to design economic policy (Prescott, 1977). However, as Cohen and Michel (1984) point out, standard control theory techniques remain useful for calculating equilibrium outcome of an appropriately specified dynamic game between a government and the far-sighted private sector. They propose two notions of time-consistency, denoted by TC_0 and TC_1. Under TC_0, the government cannot precommit at all, while under TC_1, the government is a period-by-period (or stagewise) Stackelberg leader. As an example of TC_1 in each period the government announces a consumption tax rate before the individuals consume, but the tax rate can change at the beginning of the next period. An example of TC_0 may be the inflation tax, which is not revealed before consumption. They show that TC_1 corresponds to a (stagewise) feedback Stackelberg equilibrium, in the sense defined by Başar and Olsder (1982).[1] Cohen and Michel (1984) show that in general TC_0 and TC_1 obey different HJB equations.[2]

It is useful to present a simplified version of Cohen and Michel (1984). There is a continuum of private agents, indexed by i, defined over the unit interval $[0, 1]$. Agent i has a stock $s_i(t)$ (a state variable) which he/she can partially influence by using his/her control variable $u_i(t)$. The government has a control variable $g(t)$. Define the aggregate stock

$$S(t) \equiv \int_0^1 s_i(t)di$$

The state variable $s_i(t)$ follows the law of motion

$$\dot{s}_i(t) = -As_i(t) - bu_i(t) - \gamma_i g(t), \quad A \geq 0, \ b > 0, \ \gamma_i > 0$$

---

[1]It turns out that in their linear quadratic formulation, if their "externality parameter" ε is zero, then TC_0 coincides with TC_1. See their Proposition 4.

[2]An alternative approach is to postulate that the government uses a linear policy. One then optimizes with respect to the parameters of that poicy, see, for example, d'Autume (1984) and Oudiz and Sachs (1985).

Here γ_i may differ across individuals. We may interpret that $\gamma_i g(t)$ is the tax levied on individual i. The parameter b is a measure of the effectiveness of the private control variable u_i. Individual i chooses the time path of u_i to minimize his lifetime integral of (undiscounted) quadratic losses:

$$L_i = \frac{1}{2} \int_0^\infty [Qs_i^2 + Mu_i^2 + 2\varepsilon_i u_i g] dt$$

where M and Q are parameters, $M > 0$, $Q > 0$. The externality parameter ε_i can be positive or negative. Each individual takes the time path $g(t)$ as given.[3] Let ω_i denote the co-state variable associated with s_i. The Hamiltonian for agent i is

$$H_i = \frac{1}{2} \left[Qs_i^2 + Mu_i^2 + 2\varepsilon_i u_i g \right] + \omega_i [-As_i - bu_i - \gamma_i g]$$

The necessary conditions are

$$u_i = \frac{b}{P} \omega_i - \frac{\varepsilon_i}{M} g$$
$$\dot{\omega}_i = A\omega_i - Qs_i$$

And the transversality condition is

$$\lim_{t \to \infty} \omega_i(t) = 0$$

Assuming that $\int_0^\infty [g(t)]^2 dt < \infty$, the problem of individual i has a unique solution.

Define the "normalized private sector shadow price" $\Omega(t)$ as follows

$$\Omega(t) \equiv \frac{b^2}{M} \int_0^1 \omega_i(t) di$$

and the aggregate private sector decision as

$$U(t) \equiv \int_0^1 u_i(t) di$$

---

[3]In a rational expectations equilibrium, he correctly forecasts $g(t)$ by using some Markovian rule that relates g to the aggregate state variable, S. In other words, time-consistency is imposed on government policies.

Define

$$\Gamma \equiv \int_0^1 \gamma_i d_i$$

$$E \equiv \int_0^1 \varepsilon_i di$$

$$\beta^2 \equiv \frac{Qb^2}{M}$$

$$D \equiv \Gamma - \frac{E}{M}b$$

Assume $D > 0$ for stability. We can then obtain the aggregate decision of the private sector as a function of the private sector shadow price $\Omega(t)$ and the government's control variable:

$$U(t) = \frac{1}{b}\Omega(t) - \frac{E}{M}g(t) \tag{7.1}$$

Then the time path of the aggregate stock S can be obtained by solving the following system of differential equations:

$$\dot{S} = -AS - \Omega - Dg, \quad S(0) = S_0 > 0 \tag{7.2}$$

$$\dot{\Omega} = A\Omega - \beta^2 S, \quad \lim_{t \to \infty} \Omega(t) = 0 \tag{7.3}$$

This system can be thought of as the necessary conditions of the optimization problem of a single private agent who chooses $U(t)$ to minimize the following loss function

$$L = \frac{1}{2} \int_0^\infty [QS^2 + 2EUg + MU^2]dt$$

subject to

$$\dot{S} = -AS - bU - \Gamma g$$

where $g(t)$ is taken as a given time path.

Assume that the government's objective is to minimize the loss function:

$$L_G = \frac{1}{2} \int_0^\infty (Q_G S^2 + M_G g^2)dt \tag{7.4}$$

subject to the laws of motion (7.2) and (7.3).

Remark: The reader can assume that $\varepsilon_i = \varepsilon$ for all i, $M = Q = b = \gamma_i = \Gamma = \beta = 1$ and $D = 1 - \varepsilon$ without loss of generality.

Let us consider first the "open-loop Stackelberg equilibirum": the government chooses a time path $g(.)$ to minimize Eq. (7.4) subject to Eqs. (7.2) and (7.3), by using optimal control theory. Since the initial $\Omega(0)$ is free, the government's shadow price $\mu(t)$ of this state variable (which is itself a shadow price) satisfies the transversality condition that

$$\mu(0) = 0$$

As we have pointed out in Chap. 1, this feature of OLSE gives rise to time-inconsistency: generally, $\mu(t_1) \neq 0$ for at least some $t_1 > 0$, and hence if at time t_1 the government is released from its commitment to follow the initally announced time path $g(.)$, and is allowed to replan at time t_1, it will set a new path of $g(.)$ for all $t \in [t_1, \infty]$, to reset $\mu(t_1)$ at its new optimal value, which is zero.

We conclude that the OLSE cannot be an equilibrium if agents suspect that the government is not going to keep its promises. Let us now turn to time-consistent policies.

The government follows a Markovian strategy $\phi_G(.,.)$ which determines at each (date, state) pair (t, S) the value of its policy variable $g(t)$:

$$g(t) = \phi_G(t, S)$$

Assume that under this strategy, the following system has a unique solution for all (t, x):

$$\dot{S}(\tau) = -AS(\tau) - \Omega(\tau) - D\phi_G(\tau, S(\tau)), \quad \tau \geq t, \quad x(t) = x$$

$$\dot{\Omega}(\tau) = A\Omega(\tau) - \beta^2 S(\tau), \quad \lim_{\tau \to \infty} \Omega(\tau) = 0$$

Such a strategy is called admissible. Denote the solution for the private sector shadow price by $\Omega(t, S; \phi_G)$. Then, the private sector's response is

$$U(t) = \frac{1}{b}\Omega(t, S; \phi_G) - \frac{E}{M}\phi_G(t, S) \equiv \phi_P(t, S; \phi_G) \qquad (7.5)$$

Under the admissible strategy ϕ_G, the payoff of the government is

$$V_G(t, S; \phi_G) = \frac{1}{2} \int_t^\infty [Q_G S(\tau)^2 + M_G g(\tau)^2] d\tau$$

where

$$\dot{S}(\tau) = -AS(\tau) - bU(\tau) - \Gamma g(\tau), \quad \tau \geq t, \; S(t) = S$$
$$U(\tau) = \phi_P(\tau, S(\tau); \phi_G), \quad g(\tau) = \phi_G(\tau, S(\tau))$$

An admissible strategy ϕ_G^* is said to be a time-consistent policy if $V_G(t, S; \phi_G)$ satisfies Bellman's principle of optimality. To be precise, let us state two assumptions.

Assumption P: At any t, private agents believe the current government and all future governments are going to follow strategy ϕ_G^*.

Assumption G: At any t, the current government knows that Assumption P is satisfied and that all future governments will follow ϕ_G^*.

If the model were set in discrete time, we would write

$$S(t + \Delta t) - S(t) = -AS(t) - bU(t) - \Gamma g(t)$$

and we would require ϕ_G^* to be a solution to the Bellman equation

$$V(t, S; \phi_G^*) = \min_g \left\{ \frac{(Q_G S^2 + M_G g^2)\Delta t}{2} \right.$$
$$\left. + V(t + \Delta t, S - AS - b\phi_P(t, S; \phi_G^*) - \Gamma g) \right\}$$

To obtain the corresponding HJB equation for the continuous time model, we divide both sides of the Bellman equation by Δt and then take the limit $\Delta t \to 0$. Then

$$-\frac{\partial V}{\partial t} = \min_g \left\{ \frac{(Q_G S^2 + M_G g^2)}{2} + \frac{\partial V}{\partial S}(-AS - b\phi_P(t, S; \phi_G^*) - \Gamma g) \right\}$$
$$(7.6)$$

The RHS of Eq. (7.6) is equivalent to

$$\min_g \left\{ \frac{(Q_G S^2 + M_G g^2)}{2} + \frac{\partial V}{\partial S} \left(-AS - \frac{1}{b}\Omega(t, S; \phi_G^*) + \frac{E}{M}\phi_G^*(t, S) - \Gamma g \right) \right\}$$
$$(7.7)$$

If the solution of the HJB equation yields $g = \phi_G^*(t, S)$, we say that ϕ_G^* is a time-consistent policy under assumptions P and G.

Clearly, a time-consistent policy ϕ_G^* together with the private sector strategy ϕ_P form a Nash equilibrium. In minimizing the RHS of Eq. (7.6), the government is choosing g for a given ϕ_G^*. It is not choosing the best possible ϕ_G^* taking into account that ϕ_P depends on ϕ_G^*.

In the above discussion, we have assumed that, at each point of time, the government and the private sector make their choices simultaneously. Now let us formulate a slightly different model, by assuming that in each infinitesimal period $[t, t+\Delta t]$ the private sector observes $g(t)$ before making its choice $U(t)$. Consider the following modified assumptions.

Assumption P1: At each t, private agents observe $g(t)$ before choosing $u_i(t)$, expecting all future governments to use the strategy ϕ_G^{**}.

Assumption G1: At each t, the current government knows that Assumption $P1$ is satisfied, and expects all future governments to use the strategy ϕ_G^{**}.

Since the private sector observes $g(t)$, its information at t is not (t, S) but $(t, S, g(t))$. Its action at t is no longer given by Eq. (7.5) but is instead

$$U(t) = \frac{1}{b}\Omega(t, S; \phi_G) - \frac{E}{P}g(t) \equiv \widehat{\Phi}_P(t, S, g; \phi_G) \qquad (7.8)$$

The HJB equation for the government is then

$$-\frac{\partial V}{\partial t} = \min_g \left\{ \frac{(Q_G S^2 + M_G g^2)}{2} + \frac{\partial V}{\partial S}(-AS - b\widehat{\Phi}_P(t, S, g; \phi_G^{**}) - \Gamma g) \right\} \qquad (7.9)$$

The RHS of Eq. (7.9) is equivalent to

$$-\frac{\partial V}{\partial t} = \min_g \left\{ \frac{(Q_G S^2 + M_G g^2)}{2} \right.$$
$$\left. + \frac{\partial V}{\partial S}\left(-AS - \frac{1}{b}\Omega(t, S; \phi_G^{**}) + \frac{E}{M}g - \Gamma g \right) \right\} \qquad (7.10)$$

If the solution of this HJB equation yields $g = \phi_G^{**}(t, S)$, we say that ϕ_G^{**} is a time-consistent policy under assumptions $P1$ and $G1$. This solution corresponds to the concept of "stagewise Stackelberg equilibrium" as defined by Mehlmann (1988). (Başar and Olsder (1982) used the term "feedback Stackelberg equilibrium", but the term "stagewise" is a better description.) We may say that the pair of strategies $(\phi_G^{**}, \widehat{\Phi}_P)$ is a stagewise Stackelberg equilibrium, where the government is the stagewise leader. Note that ϕ_G^{**} is a function of two variables while $\widehat{\Phi}_P$ is a function of three variables:

$$\phi_G^{**}:(t, S) \rightarrow g(t)$$

$$\widehat{\Phi}_P:(t, S, g) \rightarrow U(t)$$

It is important to bear in mind that under the stagewise Stackelberg equilibrium, the government is not seeking to influence Ω by choosing ϕ_G^{**}. In fact, ϕ_G^{**} is a fixed point, not an object of choice.

The Nash equilibrium (ϕ_G^*, ϕ_P) and the stagewise Stackelberg equilibrium $(\phi_G^{**}, \widehat{\Phi}_P)$ generally differ from each other. In the special case where $\varepsilon = 0$ (so that $E = 0$) the two equilibria coincide. Cohen and Michel (1988) show that if the strategy space is restricted to stationary linear strategies and if ε is negative, the government is *worse off* when it is a stagewise leader (as compared to the Nash equilibrium outcome). This is similar to the "disadvantageous monopoly" result obtained by Aumann (1973).

7.1.2. *Political economy of monetary and fiscal policies*

A central result of the analysis of the time-consistency problem in macroeconomic policy-making is that when the private sector believes that a benevolent government is unable to stick to a time path of its policy variables (i.e., OLSE are time-inconsistent), the time-consistent equilibrium is typically inferior to the first-best. As a reaction to this finding, many economists have argued that a welfare improvement can be achieved if some forms of partial commitments can be institutionalized. For example, central banks may be required to target the inflation rate, or nominal income, or the nominal money supply. Alternatively, the government may appoint a conservative central banker that is more concerned with inflation than the median voter.

In the Barro–Gordon (1983) model where the government tries to boost employment by creating unanticipated inflation, the long-term outcome is higher wage and price inflation. Rogoff (1985) shows that in such economies, the appointment of a conservative central banker will lead to a lower time-consistent rate of inflation. Ploeg (1995) points out that the Barro–Gordon framework relies on some form of short-run inflexibility of the labor market. He presents a model of time-inconsistent policies without labor market inflexibility. In Ploeg's model, the government has incentives to use temporary bouts of inflation and taxation to finance spending. Again, the conclusion emerges that the appointment of a conservative central banker can improve welfare.

Ploeg (1995) uses a continuous time model inspired by the discrete-time model of seigniorage formulated by Obstfeld (1989). Let $f(t)$ denote the real financial assets that the representative consumer holds at time t. The variable $f_{(t)}$ is the sum of real money balances, $m(t)$, and government bonds, $b(t)$. The real wage is assumed to be equal to output per head, and is normalized at unity, $w(t) = 1$. Let $\pi(t)$ denote the rate of inflation, and $\mu(t)$ the rate of growth of the nominal money stock. Then, by definition

$$\frac{\dot{m}}{m} = \mu - \pi \tag{7.11}$$

Define the "inflation tax payments" by[4]

$$i \equiv \pi m \tag{7.12}$$

and real seigniorage by

$$s \equiv \mu m \tag{7.13}$$

Then, we can write the following identity

$$\dot{m} = s - i \tag{7.14}$$

Let $z(t)$ be the real tax, $g(t)$ be the real government spending, and $r(t)$ be the real interest rate. The flow budget constraint of the government is

$$z + s = g + rb - \dot{b} \tag{7.15}$$

Let $c(t)$ denote real consumption. The flow budget constraint of the individual is

$$\dot{f} = rf + w - c - z - (r + \pi)m - \left[\lambda_1 \frac{z^2}{2} + \lambda_2 \frac{i^2}{2}\right] \tag{7.16}$$

The term inside the square brackets [...] represents the real costs incurred by the individual in connection with real tax payments and inflation tax payments i_t, where λ_1 and λ_2 are positive parameters. Multiply both sides of Eq. (7.16) by the integrating factor $\exp \int_0^t r(\tau)d\tau$ and integrate to obtain

---

[4]These are not taxes collected by the government, as can be seen from the government's budget constraint below; they are simply "burdens" imposed on the private sector.

the consumer's intertemporal budget constraint

$$\int_0^\infty e^{-\int_0^t r(\tau)d\tau} \left[w - c - z - (r+\pi)m - \lambda_1 \frac{z^2}{2} - \lambda_2 \frac{i^2}{2} \right] dt$$

$$= f(0) - \lim_{T\to\infty} f(T)e^{-\int_t^T r(\tau)d\tau} dt \le f(0) \tag{7.17}$$

where the inequality in (7.17) follows from the the no-Ponzi-game condition.

From Eqs. (7.15) and (7.16), we get the goods market clearance condition:

$$w - \left[\lambda_1 \frac{z^2}{2} + \lambda_2 \frac{i^2}{2} \right] = c + g \tag{7.18}$$

Ploeg (1995) assumes that the government can commit to a time path of government spending (which is exogenously given in this model) though it cannot commit to a time path of real taxes and seigniorage. Let $Q_P(t)$ denote the present value of the stream of future government spending, and let $g_P(t)$ be its annuity value

$$r(t)g_P(t) \equiv Q_P(t) \equiv \int_t^\infty g(\tau)e^{-\int_t^\tau r(\theta)d\theta}d\tau$$

Assume that

$$\lim_{t\to\infty} g_P(t) = \bar{g}_P > 0$$

where $\bar{g}_P$ is a given positive constant. Let $X(t)$ denote the stock of government commitments, defined as the sum of government debts and $Q_P(t)$

$$X(t) \equiv b(t) + Q_P(t) \equiv b(t) + r(t)g_P(t)$$

Multiply both sides of the flow government budget constraint (7.15) by the discount factor and integrate to obtain

$$\int_t^\infty [z(\tau) + s(\tau)]e^{-\int_t^\tau r(\theta)d\theta}d\tau = r(t)g_P(t)$$

$$+ \left[b(t) - \lim_{T\to\infty} b(T)e^{-\int_t^T r(\theta)d\theta}dt \right]$$

Imposing the no-Ponzi-game condition for the government, the public sector's intertemporal budget constraint is

$$\int_t^\infty [z(\tau) + s(\tau)] e^{-\int_t^\tau r(\theta)d\theta}d\tau = r(t)g_P(t) + b(t) \equiv X(t) \tag{7.19}$$

From this, we obtain the transition equation for the state variable X:

$$\dot{X}(t) = r(t)X(t) - z(t) - s(t) \qquad (7.20)$$

Assume that the utility function of the consumer is linear in the consumption of the private goods and of public services and nonlinear in real money balances:

$$U(c, g, m) = c + g + \psi(m)$$

where $\psi(.)$ is the utility of holding real money balances. Assume that $\psi(.)$ is strictly concave and attains a maximum at some value $\tilde{m} > 0$. For concreteness, let us specify

$$\psi(m) = \kappa \left[\tilde{m} \ln \left(\frac{m}{\tilde{m}} \right) + (\tilde{m} - m) \right], \quad \kappa > 0$$

Let $\delta > 0$ be the rate of utility discount. The consumer chooses the time path of consumption c and real balances m to maximize lifetime welfare:

$$W = \int_0^\infty e^{-\delta t} U(c, g, m) dt$$

subject to his intertemporal budget constraint (7.17). Let ξ be the the constant multiplier associated with the integral constraint (7.17). Pointwise differentiation with respect to $c(t)$ and $m(t)$ gives the necessary conditions for an interior solution

$$e^{-\delta t} = \xi e^{-\int_0^t r(\tau)d\tau}$$

$$e^{-\delta t} \kappa \left(\frac{\tilde{m}}{m(t)} - 1 \right) = (r(t) + \pi(t)) \xi e^{-\int_0^t r(\tau)d\tau}$$

It follows that in equilibrium, $\xi = 1$, $r(t) = \delta$, and the real demand for money balances is

$$m(t) = \frac{\kappa}{\kappa + \delta + \pi(t)} \tilde{m} \qquad (7.21)$$

Thus, a higher π implies a lower m. Substituting Eq. (7.12) into Eq. (7.21), we obtain

$$m(t) = \frac{\kappa \tilde{m} - i(t)}{\delta + \kappa} \qquad (7.22)$$

Ploeg (1995) assumes that, in equilibrium, the private sector's decision rule, reflected in Eq. (7.22), can be represented by a linear Markovian rule that relates real money demand $m(t)$ to the stock $X(t)$ of government commitments:

$$m = m(X) = \alpha_0 - \alpha_1 X, \quad \alpha_0 > 0, \quad \alpha_1 > 0 \qquad (7.23)$$

and that the government knows this rule. (Later, we will show how the parameters α_0 and α_1 are determined endogenously.) Differentiating the private sector decision rule (7.23) with respect to time, we obtain, using the transition equation for the state variable X, and the equilibrium condition $r = \delta$

$$\dot{m} = -\alpha_1 \dot{X} = -\alpha_1 \delta X + \alpha_1 z + \alpha_1 s$$

Using this equation and the identity $\dot{m} = s - i$, we obtain the private sector's expectation of i

$$i = (1 - \alpha_1)s - \alpha_1 z + \alpha_1 \delta X$$

Let us turn to the government's optimization problem. The government takes as given the private sector's Markovian decision rule (7.23). It chooses the time path of seigniorage $s(t)$ and real tax $z(t)$ to maximize the integral of its conception of social welfare, V

$$V = \int_0^\infty e^{-\delta t}[c(t) + g(t)]dt$$

subject to the transition equation (7.20), the goods market clearance condition (7.18), and the private sector's expectation of i. Then, V becomes

$$V = \int_0^\infty e^{-\delta t}\left[w - \lambda_1 \frac{z^2}{2} - \lambda_2 \frac{((1-\alpha_1)s - \alpha_1 z + \alpha_1 \delta X)^2}{2}\right]dt$$

$$\equiv \int_0^\infty e^{-\delta t}\Omega(s, z, X; \alpha_1)dt$$

The HJB equation for the government is

$$\delta V(X) = \max_{s,z}\{\Omega(s, z, X; \alpha_1) + V'(X)(\delta X - s - z)\} \qquad (7.24)$$

The first-order conditions are

$$-\lambda_1 + \alpha_1 \lambda_2((1-\alpha_1)s - \alpha_1 z + \alpha_1 \delta X) = V'(X)$$

$$(1 - \alpha_1)\lambda_2((1-\alpha_1)s - \alpha_1 z + \alpha_1 \delta X) = V'(X)$$

Now, assume that $V(X)$ is quadratic

$$V(X) = A + BX - \frac{DX^2}{2}$$

One can then solve for A, B, and D in terms of α_1 and the parameters of the model, and finally obtain the government's Markovian tax strategy

$$z = \frac{\lambda_2}{\lambda_1} \left[\frac{1 + \alpha_1}{1 + (\lambda_2/\lambda_1)} \right] \delta X \tag{7.25}$$

and its Markovian seigniorage strategy

$$s = \frac{1}{1 - \alpha_1} \left\{ \left(1 + \frac{\alpha_1 \lambda_2}{\lambda_1} \right) \left[\frac{1 + \alpha_1}{1 + (\lambda_2/\lambda_1)} \right] - \alpha_1 \right\} \delta X \tag{7.26}$$

The inflation tax is then

$$i = \left[\frac{1 + \alpha_1}{1 + (\lambda_2/\lambda_1)} \right] \delta X \tag{7.27}$$

Substituting Eq. (7.27) into Eq. (7.22) we obtain the equilibrium money holding

$$m = \frac{\kappa \tilde{m}}{\kappa + \delta} - \frac{\left[\frac{1 + \alpha_1}{1 + (\lambda_2/\lambda_1)} \right] \delta X}{\kappa + \delta} \tag{7.28}$$

Rational expectations require that Eqs. (7.28) and (7.23) are identical. This condition is satisfied if and only if

$$\alpha_0 = \frac{\kappa \tilde{m}}{\kappa + \delta}$$

and

$$\alpha_1 = \frac{\delta}{(\kappa + \delta)\beta + \kappa}$$

where β is defined as the relative importance of "shoe leather costs", that is,

$$\beta \equiv \frac{\lambda_2}{\lambda_1}$$

Substituting these values of α_0 and α_1 into the linear rule (7.23) and the government strategies (7.25) and (7.27), one obtains the government's equilibrium strategies and the private sector's equilibrium money holding strategy:

$$z = \beta \left[\frac{\kappa + \delta}{(\kappa + \delta)\beta + \kappa} \right] \delta X \tag{7.29}$$

$$i = \left[\frac{\kappa + \delta}{(\kappa + \delta)\beta + \kappa} \right] \delta X \tag{7.30}$$

$$m = \frac{\kappa \tilde{m}}{\kappa + \delta} - \frac{\delta X}{(\kappa + \delta)\beta + \kappa} \tag{7.31}$$

Commenting on Eq. (7.31), Ploeg wrote: "Households anticipate that inflation will be higher and thus hold less money balances if the outstanding government commitments are higher. The intuition is that, if commitments are high, the government requires more seigniorage and consequently private agents forecast higher inflation and hold less money balances".

Equations (7.29) and (7.30) show that inflation tax payments and conventional taxes go up and down together. What about the long-run behavior of government debts in this model? Using Eqs. (7.29) and (7.30), one can show that state variable X converges to its steady-state level $X_\infty = 0$. This means that the steady-state government debt is negative, $b_\infty = -(1/\rho)\bar{g}_P$. It also follows that the steady-state tax and seigniorage are zero.

Ploeg (1995) considers an application of the model to a heterogeneous society, where individuals have different β values (high β agents are hurt more by inflation than by conventional taxes). One can appeal to the median voter model. Let β^M be the median value of the distribution of β. Consider a permanent increase in government spending by an amount Δg. Ploeg shows that welfare cost of this increase is minimized when β is set at β^* where

$$\beta^* \approx \left(\frac{\delta + \kappa}{\kappa} \right) \beta^M > \beta^M$$

Here, β^* is interpreted as the degree of conservatism of the central banker chosen by the median voter.

7.1.3. *Dynamic games of macroeconomic policies between two countries*

A number of authors have applied dynamic game theory to the analysis of macroeconomic policies in an international context. Some authors use variants of the standard LM-IS Keynesian framework (together with a Philips curve). Miller and Salmon (1985), Oudiz and Sachs (1985), and Hallet (1984) focus on the difference between time-inconsistent and time-consistent solutions. Miller and Salmon assume that the policy variable is the interest rate. They find that policy coordination by two governments may lead to welfare losses. This result is not completely surprising. In fact it is in line with the general result that in the presence of a private sector capable of rational expectations, it can be disadvantageous to be a monopoly (Aumann, 1973; Maskin and Newbery, 1990). Oudiz and Sachs assume sluggish wage adjustments. Currie and Levine (1985) consider monetary feedback rules that are not fully optimal. Taylor (1985) considers policy coordination in a model where prices and wages are set in a staggered fashion. Using numerical simulations, Oudiz and Sachs (1985) and Taylor (1985) find that co-operation is gainful, in contrast to the numerical results of Miller and Salmon (1985) in an admittedly different model. Neck and Dockner (1995) compare OLNE and MPNE of a game of international policy making.

Turnovsky *et al.* (1988) study both the feedback Nash equilibrium and the feedback Stackelberg equilibrium in a dynamic game between two identical countries. Their model is based on the sticky-price open-economy model of Dornbusch (1976). Each country chooses the time path of its nominal money supply to influence the adjustment of the real exchange rate, defined as the price of the foreign good in terms of the home good. The objective function of each country is to minimize the discounted sum of its quadratic loss function. Let us outline the model. Call country 1 and country 2 as Home and Foreign, respectively. All Home variables are marked with an asterisk. Let Y denote Home's real output, in logarithm, measured as deviation from its natural level. Let $\mathcal{P}$ be the dollar price of Home's good, and $\mathcal{P}^*$ be the euro price of Foreign's output. Let $\mathcal{E}$ be the exchange rate (dollars per euro). In Home, the dollar price of the foreign good is $\mathcal{E}\mathcal{P}^*$. Home consumers buy both Home and Foreign goods, since they are not perfect substitutes. For them, the relative price of the Foreign good is

$$S \equiv \frac{\mathcal{E}\mathcal{P}^*}{\mathcal{P}}$$

S is called the real exchange rate. Let $P = \ln \mathcal{P}$, $P^* = \ln \mathcal{P}^*$, $E = \ln \mathcal{E}$, and $s = \ln S$. Then, s (the logarithm of the real exchange rate) is

$$s = E + P^* - P$$

Let I and I^* denote the nominal interest rates in Home and Foreign. Assume that uncovered interest parity holds, that is, Home interest rate equals Foreign interest rate plus the expected depreciation of the dollar:

$$I_t = I_t^* + (\ln \mathcal{E}_{t+1} - \ln \mathcal{E}_t) = I_t^* + E_{t+1} - E_t$$

Let $\mathcal{M}$ and $\mathcal{M}^*$ be the nominal supply of Home and Foreign, respectively. Assume that in Home, the logarithm of real money demand is a linear function of Y and I, that is, $\eta_1 Y - \eta_2 I$ where $\eta_1 > 0$ and $\eta_2 > 0$. Assume that the money market clears in each period, such that the excess demand for real money is zero:

$$\eta_1 Y_t - \eta_2 I_t - \ln\left(\frac{\mathcal{M}_t}{\mathcal{P}_t}\right) = 0$$

Home's output is demand determined, and is equal to the sum of three factors, influenced by Foreign output, the Home real interest rate, and the logarithm of the real exchange rate:

$$Y_t = \delta_1 Y_t^* - \delta_2 (I_t - P_{t+1} + P_t) + \delta_3 s_t$$

The logarithm of the consumer price index in Home is a weighted average of the logarithm of the dollar price of the Home good and the dollar price of the Foreign good:

$$\Pi_t = \alpha P_t + (1 - \alpha)(E_t + P_t^*), \quad 0 < \alpha < 1$$

Finally, the producer price is adjusted sluggishly:

$$P_{t+1} - P_t = \gamma Y_t, \quad \gamma > 0$$

A similar set of equations holds for Foreign.

Because of symmetry, at the long-run equilibrium, it holds that $s = 0$. Suppose, however, that at the beginning of the game, the (log) real exchange rate s_1 is different from zero. How would countries choose their money supply policies to react to this imbalance? Assume that the policy-maker in Home has the following loss function for period t:

$$L_t = \lambda Y_t^2 + (1 - \lambda)(\Pi_{t+1} - \Pi_t)^2, \quad 0 < \lambda < 1$$

The objective function of the policy-maker is to minimize the discounted sum of the loss function, over a fixed time horizon T, by choosing m_t (the logarithm of the real money supply):

$$\min \sum_{t=1}^{T} \beta^t L_t, \quad 0 < \beta < 1$$

The transition equation for the state variable s turns out to depend linearly on the difference in the two real money stocks:

$$s_{t+1} = cs_t + b(m_t - m_t^*)$$

where b is a negative term (derived from other parameters of the model).

In the feedback Nash equilibrium, the countries choose linear feedback policies of the form

$$m_t = \mu_\tau s_t$$

where μ_τ depends on τ, defined as the number of periods to go, $\tau \equiv T - t$. The policies, being determined by dynamic programming methods, are time-consistent. The authors find that the Nash equilibrium is unique, and use numerical simulation to illustrate the time path of various variables.[5]

An intriguing issue is what determines the initial s_1. Recall that in the original model of Dornbusch (1976), if the market is taken by surprise by a once-only permanent increase in the stock of money, the exchange rate will jump up at the initial instant and afterward follow a smooth adjustment path. In the current model, government policies are known and anticipated (from time $t = 1$). It seems, therefore, acceptable to treat s_1 as being determined by past monetary policies. The authors remark that "it would be straightforward, and lead to little change if one allowed s_1 to be determined by some trade-off of the future costs, with some initial adjustment costs".

Turning to Stackelberg equilibrium, the authors use as equilibrium concept what we call the stagewise Stackelberg leadership (see Chap. 2): in every period, a static Stackelberg game is solved.

Numerical simulations are performed to compare the feedback Nash equilibrium with the stagewise Stackelberg equilibrium, and with other possible scenarios. Starting with $s_1 = 1$, that is, an initial depreciation of home's real exchange rate relative to long-run equilibrium, what would happen in the simplest scenario, called the non-intervention case, where both countries keep their real money supplies constant? Clearly, since s_1 is higher than the long-run value $s = 0$, Home's output would be higher, and Foreign's output would be lower. With a constant real money supply, this would imply a higher home interest rate, and a lower foreign interest

---

[5]Though the authors give closed form solutions for both the feedback Nash equilibrium and the Stackelberg equilibrium, not much can be said without numerical simulations.

rate. There would be a larger expected rate of depreciation of the dollar. The higher home ouput raises the inflation rate of domestic producer price. This, together with the dollar depreciation, causes an increase in home's consumer price index.

Next, consider another simple scenario, where policy-makers only minimize their one-period loss function, resulting in a one-period Nash equilibrium.[6] Home policy-makers respond to the high s_1 by reducing its money stock to raise the interest rate, thus mitigating the output expansion. This is called "leaning against the wind". The opposite happens in Foreign. Thus, the rate of depreciation of the dollar is higher relative to the non-intervention case. The result is greater variations in the inflation rate, and higher welfare costs.

In contrast, the feedback Nash equilibrium is characterized by the opposite response. Home increases its real money supply, causing the rate of currency depreciation to decrease, and leading to a fall in the real exchange rate s. The difference between the static Nash reponse by one-period maximizers and the dynamic feedback Nash response by multiperiod maximizers is that one-period maximizers are interested only in short-term outcomes. Obviously, in the final period, the long-term maximizers become short-term maximizers and the policy rule switches sign.

Let us turn to the stagewise Stackelberg equilibrium, where Home is the leader. If there is just one period, the (static) follower reacts to the (static) leader's contraction in real money supply by an increase in its own real money supply, though by less than the full amount. Given the same initial condition (i.e., s is one unit above its long-run value), it turns out that under Home's leadership, both players are better off compared to the static Nash outcome. In contrast, when the number of periods is large, the stagewise leader's welfare costs are much reduced relative to the feedback Nash outcome, while the follower is worse off. Except for the last period, the follower tends to contract its real money supply in response to the leader's monetary contraction.

It is interesting to observe that in the feedback Nash equilibrium, in most periods Home expands its real money supply (i.e., Home is leaning with the wind) and Foreign contracts its real money supply, while in the stagewise Stackelberg equilibrium, Home (the leader) contracts its real money supply and Foreign follows suit.[7]

---

[6]This is the scenario investigated by Turnovsky and d'Orey (1986).
[7]Except for the last period.

7.2. Wealth Distribution and Endogenous
Property Rights

The long-run distribution of wealth has been a central question in growth theory. Ramsey (1928) conjectures that if households have time-additive utility functions and different rates of time preferences, then in the long run only the most patient households are owners of capital. A formal proof is given in Becker (1980), and Bewley (1982).[8] (See also Becker and Foias, 1994, where households face borrowing constraints.) However, if households have recursive utility, there exist steady states where all households own positive amounts of capital, see Lucas and Stokey (1984). Matters become more complicated if households have market power and realize that their savings will affect the rate of interest. In this case, households are players of a dynamic game of capital accumulation. Sorger (2002) investigates such a model, allowing households to be heterogeneous in terms of the degree of patience. Using the OLNE concept, Sorger shows that it is possible that in the steady state all households own positive amounts of capital. This result is generalized further by Becker (2003) and Sorger (2008), who retain the assumption that agents play open-loop strategies. Becker and Foias (2007) give sufficient conditions for the steady state of the model to be stable.

Sorger (2002) uses a discrete-time model with H infinitely lived households. Each period, household h faces the transition equation

$$k_{t+1}^h - k_t^h = w_t + f'(k_t^1 + k_t^2 + \cdots + k_t^H)k_t^h - c_t^h$$

where w is the real wage, $f(.)$ is the production function, k^h is the stock of capital owned by household h, and c^h is its consumption. The maximization problem of household h is

$$\max \sum_{t=0}^{\infty} (\beta^h)^t U^h(c_t^h)$$

where β^h is its discount factor, and $U^h(.)$ is its utility function. Each household realizes that the marginal product of capital depends on its own capital stock as well as the stocks owned by other households. Households

---

[8]In Bewley (1982), a poor household will remain poor in the steady state, that is, catching up does not take place even among agents with the same rate of time preference. Sorger (2000) proves that even with endogenous labor supplies, the rich stay rich and the poor stay poor (in relative terms). Long and Shimomura (2004a) show that the poor will eventually catch up if relative wealth appears in the utility function and the concern for relative wealth is strong enough.

are ordered according to increasing impatience, that is, household H is the most impatient one:

$$1 > \beta^1 > \beta^2 > \cdots > \beta^H > 0$$

Sorger shows that in every stationary Nash equilibrium, patient households are wealthier than impatient ones: $k^h \geq k^j$ if $h < j$, and $k^h > k^j$ whenever $k^h > 0$ and $h < j$. A numerical example shows that household 1 is not the only one to own capital in the steady state.

In the above formulation, households have market power in the capital market, but they are price takers in the labor market and the goods market. Sorger (2002) extends the model to the case where households know that the wage rate also depends on the stock of capital. Then, writing $K_t = k_t^1 + k_t^2 + \cdots + k_t^H$, the transition equation is modified as follows

$$k_{t+1}^h - k_t^h = (1/H)[f(K_t) - K_t f'(K_t)] + f'(K_t)k_t^h - c_t^h$$

For this case, numerical examples confirm the possibility that several households own capital in the steady state.

Pichler and Sorger (2009) check the robustness of the above model by considering MPNE. Using numerical methods, they find that under this concept, the steady state is unique and asymptotically stable.[9] Furthermore, the long-run distribution of wealth is typically nondegenerate. On the other hand, a major difference is that under the MPNE, the steady state depends on the utility functions. In particular, an increase in a household's intertemporal elasticity of substitution will reduce that household's steady-state capital stock and increase the capital stock of other households. Another contribution of this paper is the result about the aggregate rate of time preference. Gollier and Zeckhauser (2005) show that in a heterogeneous economy, in every Pareto-efficient allocation, the aggregate rate of time preference is a weighted average of individual rates, with the weights proportional to their tolerance of consumption fluctuations. This result no longer holds in the model of Pichler and Sorger (2009) because agents act strategically and the outcome is not Pareto efficient.

In Sorger (2002), each agent has its own capital stock, and dynamic strategic interactions occur because it is the level of the aggregate capital

---

[9]The authors apply a projection method with Chebyshev polynomials employed by Judd (1992) and Aruoba *et al.* (2006).

stock that determines the rate of return to capital. A different approach to dynamic strategic interactions is to assume that agents have common access to capital. We have discussed some models of this type in our survey of applications of dynamic games to development economics. An interesting extension of this type of model is to allow switching equilibria, as in Benhabib and Radner (1992), Benhabib and Rustichini (1996), and Tornell (1997). Benhabib and Radner (1992) present a continuous time model where agents have utility functions that are linear in consumption and have access to a common productive asset. They find an equilibrium where agents co-operate when the capital stock is low and do not co-operate when the stock is high. Benhabib and Rustichini (1996) use a discrete time model with linear technology and show that a switching equilibrium exists: the rate of exploitation switches to a lower level when the common asset reaches a certain threshold level. Thus, their economy's growth rate is increasing in the capital stock. When the linear technology is replaced by the Cobb–Douglas production function, they find that the growth rate is decreasing when the common asset is at high levels.

Tornell (1997) also considers switching equilibria in a dynamic game model between two players. In contrast to the models of Benhabib and Radner (1992) and Benhabib and Rustichini (1996), the author allows each group's share of aggregate capital to change after a switch takes place, and introduces a one-time lumpy cost at the switching time. These features generate a hump-shaped pattern of growth even though the underlying technology is linear and preferences exhibit a constant elasticity of intertemporal substitution, σ. Tornell allows three property rights regimes: common property, private property, and leader–follower. Under common property, both players have equal access to the aggregate capital stock. When one player incurs a one-time cost, it can convert the whole common property to its private property unless the other player matches. If the match takes place, the result is the private property regime, where each player has access only to its own capital stock. On the other hand, starting from the private property regime, if both players simultaneously incur each a one-time cost, the regime will revert back to the common property regime. If one player incurs a one-time cost while the other does not, the former becomes the leader and has exclusive access to the economy's capital stock. Thus, under private property, the equation of motion of player i's capital stock is

$$\dot{S}_i = AS_i - C_i, \quad i = 1, 2$$

where C_i is its consumption. Under common property

$$\dot{S} = AS - C_1 - C_2$$

where S is the aggregate capital stock. Under the leader–follower regime

$$\dot{S} = AS - C_\mathrm{L}$$

where the subscript L denotes the leader.

The solution concept used by Tornell is stationary Markov-perfect equilibrium. Thus, strategies are functions of the level of the state variables only. Trigger strategies are not allowed. Tornell (1997) restricts the maximum number of regime switches to two. This simplifying assumption allows closed form solutions, though it is somewhat arbitrary. Further, it is not clear if players know this restriction, and if they do know, how they would take it into account while formulating their strategies.

The model has some features of the pre-emption game in the industrial organization literature (Reinganum, 1981; Fudenberg and Tirole, 1985). There are some differences, however. The standard pre-emption game is a pure timing game; the only choice is when to incur a lumpy cost. In the present model, each player must choose its consumption strategy as well as its switching time. Three types of outcomes can occur in principle. It is possible that neither player chooses to incur a lumpy cost. Alternatively, both may incur their lumpy costs at exactly the same time. Finally, one player may incur the cost first and thus becomes the leader.

A key parameter in this game is σ, the elasticity of intertemporal substitution. If $\sigma \leq 1$, the common property regime may last forever. Alternatively, if the economy starts with the private property regime, this institution may also last forever. In constrast, if $\sigma > 1$, the author shows that the economy must follow a cycle: a switch from the common property regime to the private property regime and, later on, a reswitching back to common property.[10] There is no equilibrium which involves a switch to the leader–follower regime.

The dynamic game model of Tornell (1997) is able to generate results that are consistent with three stylized facts about economic growth. The first one is conditional convergence, that is, for a large set of

---

[10]Most empirical estimates indicate that $\sigma < 1$. However, Barro (2009) questions the validity of these estimates and argues that σ can be significantly greater than 1. Tornell (1996) points out that in his model the players are powerful groups, not the "representative household" for which empirical estimates of σ are obtained.

economies, the growth rate declines as income per head rises.[11] The second one is club convergence: convergence does not hold for a number of less-developed countries, some of which are trapped in poverty (Galor, 1996). The third one is that over the long run, the growth path of several major economies exhibit an inverted-U shape (Maddison, 1982; Kennedy, 1987).

A different model about the evolution of a property rights regime is proposed by Léonard and Long (2009). The authors use an overlapping-generations framework with capital accumulation. Agents predate on each other but at the same time are aware that a perfect property rights regime, if it could be costlessly enforced, would benefit everyone. Property rights enforcement uses up real resources, and the cost depends on the level of enforcement. The society must choose collectively (e.g., by voting) in each period the level of enforcement for that period, knowing that taxes must be imposed to pay for the law enforcement. The authors show that there are two interior steady-state equilibria, of which one is stable. Along the path approaching the stable steady state, the degree of enforcement increases as the stock of capital grows. The intuition is as follows. When households are poor, they do not have a great interest in a regime with a high degree of enforcement of property rights. As they grow richer through the process of capital accumulation, they become more committed to the idea of property rights protection.

Léonard and Long (2009) assume a continuum of households in the unit interval $[0, 1]$. The head of household i in period t inherits wealth $b_{it} > 0$. Let τ_t be the tax rate. The net wealth $(1 - \tau_t)b_{it}$ is allocated among consumption, c_{it}, investment, k_{it+1}, and rent-seeking, r_{it+1}. Capital depreciates completely after one period. Define the size of the economy by

$$B_t \equiv \int_0^1 b_{jt}dj$$

The total amount of capital in the economy at time $t + 1$ is

$$K_{t+1} \equiv \int_0^1 k_{jt+1}dj$$

Let z_{it+1} denote the effective influence of agent i. Because of rent-seeking, the actual amount of capital that becomes available to household i is

---

[11]See, for example, Barro (1991) and Barro and Sala-i-Martin (1995). For a survey on the endogenous growth literature in an international setting, see Long and Wong (1997).

denoted by κ_{it+1}. The size of κ_{it+1} depends on the effective influence of agent i relative to the average influence:

$$\kappa_{it+1} = \frac{z_{it+1} K_{t+1}}{\int_0^1 z_{jt+1} dj} \tag{7.32}$$

Léonard and Long (2009) assume that[12] influence can be achieved by both rent-seeking and investment, but the effectiveness of rent-seeking is smaller in a society with a higher degree of property rights enforcement. Thus, they specify the following influence-generating function

$$z_{it+1} = (k_{it+1})^{\lambda_t} (r_{it+1})^{1-\lambda_t}$$

where λ_t is the degree of property rights enforcement in period t.

At the beginning of period $t+1$, the (new) head of household i inherits the amount

$$b_{it+1} = A(\kappa_{it+1})^{\alpha} (n_{it})^{1-\alpha}$$

where n_{it} is the number of individuals in the household. For simplicity, assume $n_{it} = 1$ for all t.

Let $\beta > 0$ be the discount factor. The head of household i in period t chooses c_{it}, r_{it+1}, and k_{it+1} to maximize his welfare function

$$V_{it} = \ln(c_{it} - x_i) + \beta \ln (b_{it+1})$$

subject to its budget constraint

$$(1 - \tau_t) b_{it} = c_{it} + r_{it+1} + k_{it+1}$$

The parameter $x_i > 0$ is the minimum subsistence level of consumption.[13]
After maximization, the welfare of household i is

$$V_{it}^*(b_{it}, \lambda_t, \tau_t) = \ln \left\{ \left(\frac{1}{1+\alpha\beta} \right) [(1 - \tau_t) b_{it} - x_i] \right\} + \beta \ln A(\kappa_{it+1})^{\alpha}$$

$$= \ln \left(\frac{A^{\beta}}{1+\alpha\beta} \right) + (1+\alpha\beta) \ln[(1 - \tau_t) b_{it} - x]$$

$$+ \alpha\beta \ln \left(\frac{\alpha\beta\lambda_t}{1+\alpha\beta} \right) \tag{7.33}$$

---

[12]Gradstein (2007) assumed that $z_{it+1} = k_{it+1} r_{it+1}$.

[13]An alternative interpretation is possible. Since a bigger x_i will induce a smaller bequest b_{it+1}, x_i is a measure of selfishness or greed.

Assume that the degree of enforcement λ_t in an economy of size B_t can be achieved by government spending of $\gamma \lambda_t B_t$ units of real resources. This expenditure is financed by a tax τ_t on b_{it} for all households. A balanced budget requires $\tau_t B_t = \gamma \lambda_t B_t$. This implies that $\tau_t = \gamma \lambda_t$. Then

$$V_{it}^*(b_{it}, \lambda_t) = \ln\left(\frac{1}{1+\alpha\beta}\right) + \beta \ln A + (1+\alpha\beta)\ln[(1-\gamma\lambda_t)b_{it} - x_i]$$

$$+ \alpha\beta \ln \lambda_t + \alpha\beta \ln\left(\frac{\alpha\beta}{1+\alpha\beta}\right)$$

Household i's most preferred tax rate, denoted by λ_{it}^*, is found by maximizing V_{it}^* with respect to λ_t. The first-order condition is

$$\frac{(1+\alpha\beta)\gamma b_{it}}{(1-\gamma\lambda_{it}^*)b_{it} - x_i} = \frac{\alpha\beta}{\lambda_{it}^*} \tag{7.34}$$

Clearly the most preferred tax rate depends on wealth b_{it}:

$$\lambda_{it}^* = \frac{\alpha\beta}{\gamma(1+2\alpha\beta)}\left[1 - \frac{x_i}{b_{it}}\right] \tag{7.35}$$

Wealthier households prefer better law enforcement. Let us focus on the case of identical households. At each time period t, given b_t, they vote for the level of enforcement $\lambda_t^* = \lambda^*(b_t)$. Then

$$\lambda_t^* = \frac{\alpha\beta}{\gamma(1+2\alpha\beta)}\left[1 - \frac{x}{b_{it}}\right] \equiv \omega\left[1 - \frac{x}{b_t}\right] \equiv \lambda^*(b_t) \tag{7.36}$$

Assume $\omega \leq 1$ to ensure that $\lambda_t^* < 1$. As $b_t \to \infty$, $\lambda_t^* \to \omega$.

The evolution of the system obeys the law

$$b_{t+1} = Db_t^{-\alpha}(b_t - x)^{2\alpha} \equiv \phi(b_t) \tag{7.37}$$

where

$$D \equiv A\gamma^{-\alpha}\left(\frac{\alpha\beta}{1+2\alpha\beta}\right)^{2\alpha} > 0$$

A steady state is a fixed point b^* such that

$$b^* = \phi(b^*)$$

Then, b^* solves

$$b^* = D^{1/(1+\alpha)}[b^* - x]^{2\alpha/(1+\alpha)} \tag{7.38}$$

The LHS of Eq. (7.38) is a linear function of b^* and the RHS is a concave curve for $b^* > x$. So, there are two fixed points, b_L^* and b_H^* where $x < b_L^* < b_H^*$. Clearly b_H^* is stable, and b_L^* is unstable.

The equilibrium evolution of the property rights regime is as follows. If $b_0 \in (b_L^*, b_H^*)$, both b_t and λ_t will increase with time, converging to b_H^* and λ_H^* where

$$\lambda_H^* \equiv \omega \left(1 - \frac{x}{b_H^*} \right) < 1$$

If $b_0 \in (x, b_L^*)$, then b_t falls steadily toward x and λ_t falls steadily to zero.

7.3. Capitalists versus Workers

While Tornell (1997) and Léonard and Long (2009) focus on distributional conflicts between symmetric groups, there is a literature that deals with dynamic distributional conflicts between capitalists and workers.[14] The first model of this kind is Lancaster (1973). The model is set in continuous time, with a fixed terminal time T. The stock of capital is $k(t)$. Only capitalists own capital, while workers provide labor. Output is $Ak(t)$ where A is a positive constant. Labor is not a limiting factor. At each t, workers as a group choose a fraction $u_1(t)$ of output to consume, where $u_1 \in [c, b]$ and $0 < c < b < 1$. Here, c and b are exogenous constants. What is left for the capitalists is the "net output" $Ak(t)[1 - u_1(t)]$. The capitalists, as a group, possess the control variable $u_2(t) \in [0, 1]$, which is the fraction of the net output to be invested. Capitalists consume $Ak(t)[1 - u_1(t)][1 - u_2(t)]$. The objective functions of the two players are, respectively

$$J_1 = \int_0^T Ak(t)[1 - u_1(t)]dt$$

and

$$J_2 = \int_0^T Ak(t)[1 - u_1(t)][1 - u_2(t)]dt$$

---

[14]We should also mention a different set of models about conflicts. Simaan and Cruz (1975) model a dynamic game of arms race between two countries. Ploeg and de Zeeuw (1990) show that the MPNE leads to lower stocks of arms and higher welfare than the OLNE. Amegashie and Runkel (2008) offer a dynamic game model of conflicts in which both parties want to exact revenge.

The transition equation is

$$\dot{k}(t) = Ak(t)[1 - u_1(t)]u_2(t)$$

This game is a state-linear game (in the sense defined in Chap. 7 of Dockner *et al.*, 2000). Observe that from the necessary conditions obtained using the maximum principle, the control variable at any time t depends only on the shadow price at t, and the equation of motion of the shadow price is independent of the state variable. Therefore, as proved in Dockner *et al.* (2000), for such models the OLNE and the MPNE coincide. Lancaster shows that the Nash equilibrium of this game is inefficient (as compared with the socially optimal program that maximizes $J_1 + J_2$). In the Nash equilibrium there is an accumulation phase, followed by a phase of zero accumulation. This pattern also holds in the socially optimal program, but in the Nash equilibrium, the accumulation phase stops too early.

Hoel (1978b) generalizes the Lancaster model by allowing (i) concave utility and production functions and (ii) discounting. He considers only the OLNE. (With a strictly concave production function, the OLNE no longer coincides with MPNE.)

Pohjola (1983) derives the OLSE for the Lancaster model. He finds that both in the case where the workers are the leader and in case where the capitalists lead, there is more capital accumulation than in the Nash equilibrium. Under either Stackelberg equilibrium, both players are better off as compared to their Nash payoffs. However, when the payoffs under the two Stackelberg equilibria are compared, the workers prefer the capitalists to lead, while the capitalists prefer the workers to lead. Pohjola concludes that "the Stackelberg game is in a stalemate: neither player wants to act as the leader".

As we have noted in Chap. 1, OLSE are in general time-inconsistent. To overcome this objection, Başar *et al.* (1985a, 1985b) investigate the feedback Stackelberg equilibria of the Lancaster model, using a discrete time formulation. The authors specifically use the concept of a "*stagewise asymmetric mode of play*" according to which "one of the players announces his action before the other player does, at each stage" (p. 106). The authors refer to the corresponding equilibrium as the feedback Stackelberg solution.[15] They begin with a "two-period" model, in which all actions

---

[15]The more descriptive terminology "stagewise Stackelberg equilibrium" is used by Mehlmann (1988).

take place in period 1: in period 2, the two players only receive their "scrap value" of the game, denoted by $V_1(K)$ and $V_2(K)$, respectively, where K is the capital stock at the end of period 1. They modify the Lancaster model by assuming that when the workers demand u_1, the output of the economy shrinks from Ak to $Ak\psi(u_1)$, where $1 \geq \psi(u_1) > 0$ is the shrinkage factor. This shrinkage represents the social cost of strikes. It is assumed that for all $u_1 \in (0,1)$

$$\psi'(u_1) < 0, \quad u_1\psi'(u_1) + \psi(u_1) > 0, \quad \psi''(u_1) < 0$$

The workers' consumption is then $Ak\psi(u_1)u_1$. The amount of output left for the capitalists is then $Ak\psi(u_1)[1 - u_1]$. If they invest a fraction u_2 of this "net output", then the end-of-period capital stock is $K = Ak\psi(u_1)[1-u_1]u_2$, and the consumption of the capitalists is $Ak\psi(u_1)[1-u_1][1-u_2]$. As in Lancaster's model, it is assumed that $0 < c \leq u_1 \leq b < 1$ and $0 \leq u_2 \leq 1$.

Başar *et al.* (1985a) specify the following scrap values for the workers and the capitalists, respectively

$$V_1(K) = \beta_1 K$$
$$V_2(K) = \beta_2 K$$

where β_1 and β_2 are positive and exogenous constants. The case where both β_1 and β_2 are greater than unity is admitted. The objective functions of the players are

$$J_1(k, u_1, u_2) = Ak\psi(u_1)u_1 + V_1(K) \tag{7.39}$$

$$J_2(k, u_1, u_2) = Ak\psi(u_1)[1 - u_1][1 - u_2] + V_2(K) \tag{7.40}$$

For this "two-period" game, it can be verified that the decision of the capitalists is independent of the value of u_1 that the workers choose. Therefore, the Stackelberg equilibrium when the workers lead is identical to the Nash equilibrium. Turning to the Stackelberg equilibrium where the the capitalists lead, one can show that it yields better payoffs for both players (as compared to their Nash equilibrium payoffs). The value functions turn out to be linear for both players, both under the Nash equilibrium and the capitalist-led Stackelberg equilibrium.

The authors extend their analysis to the case of T periods. The results are much the same as in the "two-period" game. The method of analysis is based on dynamic programming, using the approach developed by Başar and Haurie (1984). Finally, by letting the time interval shrink to zero, a

continuous-time version of the model is obtained, and the basic results stay
unchanged. The stagewise Stackelberg solution where the capitalists lead
dominates the feedback Nash solution. (The latter is identical to the open-
loop Nash solution, because the game is state linear, and also identical to
the feedback Stackelberg solution where the workers lead.)

Shimomura (1991) revisits the differential game of capitalism by
allowing the production function to be strictly concave in the capital-labor
ratio. Concerning utility functions, Shimomura assumes constant absolute
risk aversion. The utility function of the representative worker is

$$U^1(W) = 1 - \exp(-\alpha_w W), \quad \alpha_w > 0$$

and that of the representative capitalist is

$$U^2(C) = 1 - \exp(-\alpha_c C), \quad \alpha_c > 0$$

The transition equation is

$$\dot{k} = f(k) - nk - C - W$$

Taking into account the transition equation, the workers collectively
maximize

$$\int_0^\infty e^{-\rho t}[1 - \exp(-\alpha_w W)]dt$$

while the capitalists collectively maximize

$$\int_0^\infty e^{-\rho t}[1 - \exp(-\alpha_c C)]dt$$

Even though α_c and α_w can be different, this formulation is symmetrical,
in the sense that there is no real difference in the optimization problems
facing capitalists and workers. (Compare with the asymmetric formulation
of Başar et al., 1985a, as reflected in Eqs. (7.39) and (7.40) which describe
two really different objective functions.) One might as well think of two
groups, called group 1 and group 2. Let one group be the stagewise leader
and the other group be the follower.[16] Shimomura shows that the feedback
Stackelberg equilibrium is identical to the feedback Nash equilibrium. This
result is quite different from that of Başar et al. (1985a), essentially because
Shimomura has chosen a completely symmetric formulation.

---

[16]Shimomura did not use the word "stagewise".

Shimomura shows the existence of a feedback Nash equilibrium, and confirms that it is inefficient. In this equilibrium, when the capital stock is low, both groups refrain from consumption. This is because the marginal utility of consumption is finite at all consumption levels, due to the constant-absolute risk-aversion assumption.

Krawzyck and Shimomura (2003) modify the model of Shimomura (1991) by assuming that the capitalists enjoy not only consumption but also wealth. Their utility function is

$$U^2(C, K) = BK^b + DC^b, \quad 0 < b < 1$$

The ratio B/D is interpreted as the degree of greed of the capitalist class. The workers have the utility function

$$U^1(W) = W^b, \quad 0 < b < 1$$

Net output is AK, where $A > 1$ is a constant.

The equation of motion is

$$K_{t+1} = AK_t - C_t - W_t, \quad A > 1$$

Let $\beta \in (0, 1)$ be the common discount factor. Starting at time τ when the stock of capital is K_τ, the welfare of the representative infinitely-lived worker is

$$J^1(\tau, K_\tau; W, C) = \sum_{t=\tau}^{\infty} \beta^t (W_t)^b$$

and that of the representative infinitely lived capitalist is

$$J^2(\tau, K_\tau; W, C) = \sum_{t=\tau}^{\infty} \beta^t [B(K_t)^b + D(C_t)^b]$$

Suppose the capitalists have the consumption strategy

$$C_t = \phi^2(K_t)$$

and the workers have the wage-demand strategy

$$W_t = \phi^1(K_t)$$

The Bellman equations for the two players are

$$V^1(K) = \max_W\{W^b + \beta V^1(AK - \phi^2(K) - W)\}$$

$$V^2(K) = \max_C\{W^b + \beta V^2(AK - C - \phi^1(K))\}$$

The authors show that if $b < 1/2$ and $A^b < 1/\beta < A$, there exists a unique feedback Nash equilibrium where the players use linear Markovian strategies. When the equilibrium strategies are played, the resulting growth rate of the economy is a constant μ. A greater degree of greed B/D will result in a higher growth rate.

Consider now a social planner that wishes to maximize a weighted sum of welfare, $(1-\alpha)J^1(\tau, K_\tau; W, C) + \alpha J^2(\tau, K_\tau; W, C)$, where α is the weight the planner attaches to the welfare of the capitalists. The authors suggest that α might be thought of as the "bargaining power" of the capitalists. They find that the planner will select decision rules for W and C that are linear in the state variable K if and only if $A^b < 1/\beta$:

$$W = \underline{w}K$$

$$C = \underline{c}K$$

where it is understood that $\underline{w}$ and $\underline{c}$ depend on α. A higher α will result in a higher growth rate.

The authors then ask if the planner's allocation (which depends on α) can be supported by a collusive equilibrium using trigger strategies, as in Green and Porter (1984). Following Haurie and Towinski (1990), and Haurie *et al.* (1994), Krawzyck and Shimomura consider an extended state pair (K, Y) where Y is a binary index which can take on the value of zero or one. If $Y = 1$, the system is in the co-operative mode. When Y is switched to zero, this indicates that at least one of the players has "defected". The authors find that if α lies within a certain interval $[\underline{\alpha}, \overline{\alpha}]$, then the planner's allocation can be supported by a subgame-perfect equilibrium using trigger strategies.

From this model, Krawzyck and Shimomura argue that countries with the same technology and class-specific utility functions may experience different growth rates because they have different values for α (which reflects their institutional or organizational structures).

REFERENCES

Açikgöz, O and H Benchekroun (2005). Detrimental Future International Agreements, Typescript, McGill University.

Alvarez-Cuadrado, F and NV Long (2009). A mixed Bentham-Rawls criterion for intergenerational equity: Theory and implications. *Journal of Environmental Economics and Management*, 58(2), 154–168.

Amegashie, JA and E Kutsoati (2007). Non-intervention in intra-state conflicts. *European Journal of Political Economy*, 23, 754–767.

Amegashie, JA and M Runkel (2008). The Paradoxes of Revenge and Conflicts, Typescript, University of Guelph.

Amir, R (1996). Continuous stochastic games of capital accumulation with convex transitions. *Games and Economic Behavior*, 15, 111–131.

Antoniadou, E, C Koulovatianos and L Mirman (2007). Strategic Exploitation of Common-Property Resource under Uncertainty, Working Paper 0703, Department of Economics, University of Vienna.

Aruoba, B, J Fernandez-Villaverde and J Rubio-Ramirez (2006). Comparing solution methods for dynamic equilibrium economies. *Journal of Economic Dynamics and Control*, 30, 2477–2508.

Asheim, GB (1988). Rawlsian intergenerational justice as a Markov-perfect equilibrium in a resource technology. *Review of Economic Studies*, 55(3), 469–483.

Aumann, RJ (1973). Disadvantageous monopolies. *Journal of Economic Theory*, 6, 1–11.

Ausubel, LM and RJ Deneckere (1987). One is almost enough for monopoly. *Rand Journal of Economics*, 18, 255–274.

Bagnoli, M and BL Lipman (1989). Provision of public goods: Fully implementing the core through private contribution. *Review of Economic Studies*, 56, 583–601.

Bahn, O and A Haurie (2008). A Class of Games with Coupled Constraints to Model International GHG Emission Agreements. *Les Cahiers du GERAD*, G-2008-73.

Bahn, O, A Haurie and R Malhamé (2008). A stochastic control model for optimal timing of climate policies. *Automatica*, 44, 1545–1558.

Baksi, S and NV Long (2009). Recycling with endogenous consumer participation. *Australian Economic Papers*, 48(4), 281–295.

Balboa, O, R Driskill and A Horowitz (2007). The time-consistent optimal export policy, market structure, and time-non-separable preferences. *Asia-Pacific Journal of Accounting & Economics*, 14(3), 293–314.

Barrett, S. (1994). Self-enforcing international environmental agreements. *Oxford Economic Papers*, 46, 878–894.

Barrett, S. (1997). Toward a theory of international environmental cooperation. In *New Directions in the Economic Theory of the Environment*, C Carraro and D Siniscalco (eds.), pp. 239–280. Cambridge, UK: Cambridge University Press.

Barrett, S (2003). *Environment & Statecrafts. The Strategy of Environmental Treaty-Making*. Oxford: Oxford University Press.

Barro, R (1973). The control of politicians: An economic model. *Public Choice*, 14, 19–42.

Barro, R (1983). Inflationary finance under rules and discretion. *Canadian Journal of Economics*, 41, 1–16.

Barro, R (1991). Economic growth in a cross section of countries. *Quarterly Journal of Economics*, 106, 407–443.

Barro, R (2009). Rare disasters, asset prices, and welfare costs. *American Economic Review*, 99(1), 243–264.

Barro, R and DB Gordon (1983). Rules, discretion, and reputation in a model of monetary policy. *Journal of Monetary Economics*, 12, 101–121.

Barro, R and X Sala-i-Martin (1995). *Economic Growth*. New York: McGraw-Hill.

Başar, T (ed.) *Dynamic Games and Applications in Economics*. Berlin and New York: Springer.

Başar, T and A Haurie (1984). Feedback equilibrium in differential games with structural and modal uncertainty. In *Advances in Large Scale Systems*. J Curz (ed.), Vol. 1, pp. 163–201.

Başar, T, A Haurie and G Ricci (1985a). On the dominance of capitalists leadership in a 'feedback Stackelberg' solution of a differential game model of capitalism. *Journal of Economic Dynamics and Control*, 9, 101–125.

Başar, T, A Haurie and G Ricci (1985b). Errata. *Journal of Economic Dynamics and Control*, 9, 493.

Başar, T and GJ Olsder (1982). *Dynamic Noncooperative Game Theory*. London: Academic Press.

Başar, T and GJ Olsder (1995). *Dynamic Noncooperative Game Theory*, 2nd Ed. London: Academic Press.

Başar, T and H Selbuz (1979). Closed-loop Stackelberg strategies with applications in the optimal control of multilevel systems. *Institute of Electronics and Electronics Engineers Transactions on Automatic Control*, AC-24, 166–170.

Batabyal, AA (1996a). Consistency and optimality in a dynamic game of pollution control I: Competition. *Environmental and Resource Economics*, 8, 205–220.

Batabyal, AA (1996b). Consistency and optimality in a dynamic game of pollution control II: Monopoly. *Environmental and Resource Economics*, 8, 315–330.

Becker, R (1980). On the long-run steady state in a simple dynamic model of equilibrium with heterogeneous households. *Quarterly Journal of Economics*, 95, 375–382.

Becker, R (2003). Stationary strategic Ramsey equilibrium, Working Paper, Indiana University.

Becker, R and C Foias (1987). A characterization of Ramsey equilibrium. *Journal of Economic Theory*, 41, 173–184.

Becker, R and C Foias (2007). Strategic Ramsey equilibrium dynamics. *Journal of Mathematical Economics*, 43, 318–346.

Belleflamme, P and J Hindricks (2005). Yardstick competition and political agency problems. *Social Choice and Welfare*, 24, 155–159.

Benchekroun, H (2003a). The closed-loop effect and the profitability of horizontal mergers. *Canadian Journal of Economics*, 36(3), 546–565.

Benchekroun, H (2003b). Unilateral production restrictions in a dynamic duopoly. *Journal of Economic Theory*, 111(2), 214–39.

Benchekroun, H (2007). A unifying differential game of advertising and promotions. *International Game Theory Review*, 9(2), 183–197.

Benchekroun, H (2008). Comparative dynamics in a productive asset oligopoly. *Journal of Economic Theory*, 138, 237–261.

Benchekroun, H and G Gaudet (2003). On the profitability of production perturbation in a dynamic natural resource oligopoly. *Journal of Economic Dynamics and Control*, 27(7), 1237–1252.

Benchekroun, H, G Gaudet and NV Long (2006). Temporary natural resource cartels. *Journal of Environmental Economics and Management*, 52(3), 663–674.

Benchekroun, H, A Halsema and C Withagen (2009). On non-renewable resource oligopolies: The asymmetric case. *Journal of Economic Dynamics and Control*, 33, 1867–1879.

Benchekroun, H, A Halsema and C Withagen (2010). When additional resource stocks reduce welfare. *Journal of Environmental Economics and Management*, 59(1), 109–144.

Benchekroun, H and NV Long (1998). Efficiency-inducing taxation for polluting oligopolists. *Journal of Public Economics*, 70, 325–342.

Benchekroun, H and NV Long (2002a). Transboundary fishery: A differential game model. *Economica*, 69, 207–221.

Benchekroun, H and NV Long (2002b). On the multiplicity of efficiency-inducing tax rules. *Economics Letters*, 76, 331–336.

Benchekroun, H and NV Long (2006). The curse of windfall gains in a non-renewable resource oligopoly. *Australian Economic Papers*, 45(2), 99–105.

Benchekroun, H and NV Long (2008a). The build up of cooperative behavior among non-cooperative agents. *Journal of Economic Behavior and Organization*, 67(1), 293–252.

Benchekroun, H and NV Long (2008b). A class of performance-based subsidy rules. *Japanese Economic Review*, 59(4), 381–400.

Benchekroun, H and C Withagen (2008). Non-renewable resource oligopoly and the cartel-fringe game. *Cireq Working Paper* 14-2008, Department of Economics, Université de Montréal.

Benhabib, J and G Ferri (1987). Bargaining and the evolution of cooperation in a dynamic game. *Economics Letters*, 24, 107–111.

Benhabib, J and R Radner (1992). The joint exploitation of a productive asset: A game theoretic approach. *Economic Theory*, 2, 155–190.

Benhabib, J and A Rustichini (1994). A note on a new class of solutions to dynamic programming problems arising in economic growth. *Journal of Economic Dynamic and Control*, 18, 807–813.

Benhabib, J and A Rustichini (1996). Social conflict and growth. *Journal of Economic Growth*, 1, 125–142.

Berg, J, J Dickhaut and K McKabe (1995). Trust, reciprocity, and social history. *Games and Economic Behavior*, 10, 122–142.

Bergstrom, T (1982). On capturing oil rents with national excise tax. *American Economic Review*, 71, 194–201.

Bergstrom, T, C Blume and H Varian (1986). On the private provision of public goods. *Journal of Public Economics*, 29, 25–49.

Bergstrom, T, C Blume and H Varian (1992). Uniqueness of Nash equilibrium in the private provision of public goods: An improved proof. *Journal of Public Economics*, 49, 391–392.

Bergstrom, T, J Cross and R Porter (1981). Efficiency-inducing taxation for a monopolistically supplied depletable resources. *Journal of Public Economics*, 15, 23–32.

Bernard, A, A Haurie, M Vielle and L Viguier (2008). A two-level dynamic game of carbon emission trading between Russia, China, and Annex B countries. *Journal of Economic Dynamics and Control*, 32, 1830–1856.

Bernard, A, S Paltsev, J Reilly, M Vielle and L Viguier (2003). Russia's Role in the Kyoto Protocol, Report 98, MIT Joint Program on the Science and Policy of Global Change, Cambridge, MA.

Bernard, A and M Vielle (1998). La structure du modèle GEMINI-E3. *Economie et Prévision*, 5(136), 19–32.

Bernard, A and M Vielle (2000). Comment allouer un coût global d'environnement entre pays: permis négociables versus taxes ou permis nègociables et taxes? *Economie Internationale*, 82, 103–135.

Bernard, A and M Vielle (2003). Measuring the welfare cost of climate change policies: A comparative assessment based on the computable general equilibrium model GEMINI-E3. *Environmental Modeling and Assessment*, 8(3), 199–217.

Bernheim, BD and D Ray (1983). Altruistic growth economies: I. Existence of bequest equilibria, Technical Report Number 419, Institute for Mathematical Studies in the Social Sciences, Standford University.

Bernheim, BD and D Ray (1987). Growth with altruism. *Review of Economic Studies*, 54, 227–243.

Besley, T and A Case (1995). Incumbent behavior: Vote-seeking, tax-setting, and yardstick competition. *American Economic Review*, 85, 25–45.

Besley, T and M Smart (2001). Globalization and Electoral Accountability, typescript, University of Toronto.

Bewley, T (1982). An integration of equilibrium theory and turnpike theory. *Journal of Mathematical Economics*, 10, 233–267.

Boadway, R, P Pestiau and D Wildasin (1989a). Non-cooperative behavior and efficient provision of public goods. *Public Finance*, 44, 1–7.

Boadway, R, P Pestiau and D Wildasin (1989a). Tax-transfer policies and the voluntary provision of public goods. *Journal of Public Economics*, 39, 157–176.

Bohringer, C (2001). Climate politics from Kyoto to Bonn: From little to nothing? *Discussion Paper* 01-49, ZEW, Mannheim, Germany.

Bolle, F (1980). The efficient use of a non-renewable common pool resource is possible but unlikely. *Zeitschrift für Nationalökonomie*, 40, 391–397.

Brander, J and S Djajic (1983). Rent extracting tariffs and the management of exhaustible resources. *Canadian Journal of Economics*, 16, 288–298.

Brander, J and B Spencer (1985). Export subsidies and international market share rivalry. *Journal of International Economics*, 11, 1–14.

Brander, J and SM Taylor (1997). International trade and open access renewable resources: The small open economy case. *Canadian Journal of Economics*, 30(3), 526–552.

Breton, M, G Martín-Herrán and G Zaccour (2006). Equilibrium investment strategies in foreign environmental projects. *Journal of Optimization Theory and Applications*, 130(1), 23–40.

Breton, M, L Sbraga and G Zaccour (2008). Dynamic Model for International Environmental Agreements. *Nota di Lavoro* 33.2008, Fondazione Eni Enrico Mattei.

Breton, M, G Zaccour and M Zahaf (2005). A differential game of joint implementation of environmental projects. *Automatica*, 41(10), 1737–1749.

Breton, M, G Zaccour and M Zahaf (2006). A game-theoretic formulation of joint implementation of environmental projects. *European Journal of Operational Research*, 168(1), 221–239.

Brock, WA (1977). A polluted golden age. In *Economics of Natural and Environmental Resources*, V Smith (ed.), New York: Gordon and Breach.

Buchner, B, C Carraro and I Cersosimo (2002). On the consequence of the US withdrawal from the Kyoto/Bonn Protocol. *Climate Policy*, 2, 273–292.

Buiter, W (1983). Optimal and Time-Consistent Policies in Continuous Time, Rational Expectations Models, NBER Technical Working Paper 29.

Bulow, J, J Geanakoplos and P Klemperer (1985). Multimarket oligopoly: Strategic substitutes and complements. *Journal of Political Economy*, 93, 488–511.

Cabo, F, E Escudero and G Martín-Herrán (2006). A time-consistent agreement in an interregional differential game on pollution and trade. *International game theory review*, 8(3), 369–393.

Cabo, F and G Martín-Herrán (2006). North-South transfer vs. biodiversity conservation: A trade differential game. *Annals of Regional Science*, 40, 249–278.

Calvo, G (1978a). On the time consistency of optimal policy in a monetary economy. *Econometrica*, 6, 1411–1428.

Calvo, G (1978b). Some notes on time inconsistency and Rawls' maximin criterion. *Review of Economic Studies*, 45, 97–102.

Calzolari, G and L Lambertini (2006). Tariffs vs quotas in a trade model with capital accumulation. *Review of International Economics*, 14, 632–644.

Calzolari, G and L Lambertini (2007). Exports restraints in a model of trade with capital accumulation. *Journal of Economic Dynamics and Control*, 31, 3822–3842.

Carlson, DA and A Haurie (2000). Infinite horizon dynamic games with coupled state constraint. *Annals of the International Society of Dynamic Games*, 5, 196–212.

Carpenter, JP (2004). When in Rome: Conformity and the provision of public goods. *Journal of Socio-Economics*, 33(4), 395–408.

Carpenter, JP and P Matthews (2005). No switch backs: Rethinking aspiration-based dynamics in the ultimatum game. *Theory and Decision*, 58(4), 351–385.

Carpenter, JP, AG Daniere and LM Takahashi (2004). Cooperation, trust, and social capital in Southeast Asian urban slums. *Journal of Economic Behavior & Organization*, 55, 533–551.

Carraro, C and D Siniscalco (1993). Strategies of international protection of the environment. *Journal of Public Economics*, 52, 309–328.

Cave, J (1987). Long-term competition in a dynamic game: The cold fish war. *Rand Journal of Economics*, 18, 596–610.

Cellini, R and L Lambertini (1998). A dynamic model of differentiated oligopoly with capital accumulation. *Journal of Economic Theory*, 83, 145–155.

Chamley, C (1986). Optimal taxation of capital income in general equilibrium with infinite lives. *Econometrica*, 54, 607–622.

Chandler, P and H Tulkens (1995). A core theoretic solution for the design of cooperative agreements on transfrontier pollution. *International Tax and Public Finance*, 2, 279–293.

Chang, F-R (2004). *Stochastic Optimization in Continuous Time*. Cambridge University Press.

Cheng, Leonard LK (1987). Optimal trade and technology policies: Dynamic linkages. *International Economic Review*, 28, 757–776.

Chiarella, C (1980). Trade between resource-poor and resource-rich economies as a differential game, Essay 19. In *Exhaustible Resources, Optimality and Trade*, MC Kemp and NV Long (eds.), pp. 219–246, Ansterdam: North Holland.

Chiarella, C, MC Kemp, NV Long and K Okuguchi (1984). On the economics of international fisheries. *International Economic Review*, 25(1), 85–92.

Chichilnisky, G (1996). An axiomatic approach to sustainable development. *Social Choice and Welfare*, 13(3), 231–257.

Choi, JP (1991). Dynamic R&D competition under 'hazard rate' uncertainty. *Rand Journal of Economics*, 22(4), 596–610.

Chou, S and NV Long (2009). Optimal tariffs on exhaustible resources in the presence of cartel behavior. *Asia-Pacific Journal of Accounting and Economics*, 16(3), 239–254.

Clark, CW (1990). *Mathematical Bioeconomics: The Optimal Management of Renewable Resources*, 2nd Ed. New York: Wiley.

Clark, CW and GR Munro (1975). The economics of fishing and modern capital theory. *Journal of Environmental Economics and Management*, 2, 92–106.

Clemhout, S, G Leitmann and HY Wan, Jr. (1973). A differential game model of oligopoly. *Journal of Cybernetics*, 3(1), 24–30.

Clemhout, S and HY Wan, Jr. (1985). Dynamic common-property resources and environmental problems. *Journal of Optimization Theory and Applications*, 46, 471–481.

Clemhout, S and HY Wan, Jr. (1993). The non-uniqueness of Markovian strategy equilibrium: The case of continuous-time models for non-renewable resources. In *Annals of the International Society of Dynamic Games: Advances in Dynamic Games and Applications*, T Başar and A Haurie (eds.). Boston: Birkhauser.

Clemhout, S and HY Wan, Jr. (1994). Differential games: Economic applications. In *Handbook of Game Theory with Economic Applications II*, RJ Aumann and S Hart (eds.), pp. 801–825, Amsterdam: Elsevier.

Clemhout, S and HY Wan, Jr. (1996). Endogenous growth as a dynamic game. In *Annals of the International Society of Dynamic Games: New Trends in Dynamic Games and Applications*, GJ Olsder (ed.). Boston: Birkhauser.

Coase, R (1972). Durability and monopoly. *Journal of Law and Economics*, 15, 143–149.

Cohen, D and P Michel (1988). How should control theory be used to calculate a time-consistent government policy? *Review of Economic Studies*, 55, 263–274.

Cole, H, G Mailath and A Postlewaite (1992). Social norms, savings behavior, and growth. *Journal of Political Economy*, 100, 1092–1125.

Copeland, BR and SM Taylor (2009). Trade, tragedy, and the commons. *American Economic Review*, 99(3), 725–749.

Cornes, RC (1993). Dyke maintenance and other stories: Some neglected types of public goods. *Quarterly Journal of Economics*, 108(1), 259–271.

Cornes, RC, NV Long and K Shimomura (2001). Drugs and pests: Intertemporal production externalities. *Japan and the World Economy*, 13(3), 255–278.

Cornes, RC and T Sandler (1984). The theory of public goods: Non-Nash behavior. *Journal of Public Economics*, 23, 367–379.

Cornes, RC and T Sandler (1985). On the Consistency of Conjectures with Public Goods. *Journal of Public Economics*, 27, 125–129.

Cornes, RC and T Sandler (1996). *The Theory of Externalities, Public Goods and Club Goods*. Cambridge, UK: Cambridge University Press.

Crabbé, P and NV Long (1993). Entry deterrence and overexploitation of the fishery. *Journal of Economic Dynamics and Control*, 17(4), 679–704.

Crawford, V, J Sobel and I Takahashi (1980). Bargaining, Strategic Reserves and International Trade in Exhaustible Resources, University of California -San Diego, Discussion Paper No. 80-30.

Criqui, P (1996). POLES 2.2. Technical Report, JOULE II Programme. *European Commission DP XVII*, Science Research Development, Brussel, Belgium.

Currie, D and P Levine (1985). Macroeconomic policy design in an interdependent world. In *International Economic Policy Coordination*, WH Buiter and RC Marston (eds.). New York: Cambridge University Press.

Dasgupta, P (1974a). Some problems arising from professor Rawls' conception of distributive justice. *Theory and Decision*, 4, 325–344.

Dasgupta, P (1974b). On some alternative criteria for justice between generations. *Journal of Public Economics*, 3, 405–423.

Dasgupta, P, R Gilbert and JE Stiglitz (1983). Strategic considerations in invention and innovation: The case of natural resources. *Econometrica*, 51, 1439–1448.

Dasgupta, P and G Heal (1979). *Economic Theory and Exhaustible Resources*. Cambridge, UK: Cambridge University Press.

Dasgupta, P and JE Stiglitz (1981). Resource depletion under technological uncertainty. *Econometrica*, 49, 85–104.

Datta, M and L Mirman (1999). Externalities, market power and resource extraction. *Journal of Environmental Economics and Management*, 37, 233–255.

Davidson, R (1978). Optimal depletion of an exhaustible resource with research and development toward an alternative technology. *Review of Economic Studies*, 45, 335–367.

d' Autume, A (1984). Closed-loop Dynamic Games and the Time-Consistency Problem. *Working Paper* 84–11, Brown University.

de Zeeuw, A (1998). International dynamic pollution control. In *Game Theory and the Environment*, N Hanley and H Folmer (eds.), Cheltenham, UK: Edward Elgar, pp. 273–354.

de Zeeuw, A (2008). Dynamic effects on the stability of international environmental agreements. *Journal of Environmental Economics and Management*, 55, 163–174.

de Zeeuw, A and F van der Ploeg (1991). Difference games and policy evaluation: A conceptual framework. *Oxford Economic Papers*, 43, 612–636.

Dockner, E (1985). Local stability analysis in optimal control problems with two state variables. In *Optimal Control Theory and Economic Analysis 2*, G Feichtinger (ed.), Amsterdam: North Holland, pp. 89–103.

Dockner, E (1992). A dynamic theory of conjectural variations. *Journal of Industrial Economics*, 40, 377–395.

Dockner, E and G Feichtinger (1991). On the optimality of limit cycles in dynamic economic systems. *Journal of Economics*, 53, 31–50.

Dockner, E, G Feichtinger and A Mehlmann (1989). Non-cooperative solutions for a differential game model of fishery. *Journal of Economic Dynamics and Control*, 13, 11–20.

Dockner, E and A Gaunerdorfer (2001). On the profitability of horizontal mergers in industries with dynamic competition. *Japan and the World Economy*, 13(3), 195–216.

Dockner, E and A Haug (1990). Tariffs and quotas under dynamic duopolistic competition. *Journal of International Economics*, 29, 147–160.

Dockner, E and A Haug (1991). The closed-loop motive for voluntary export restraints. *Canadian Journal of Economics*, 24(3), 679–685.

Dockner, E, S Jørgensen, NV Long and G Sorger (2000). *Differential Games in Economics and Management Science.* Cambridge, UK: Cambridge University Press.

Dockner, E and V Kaitala (1989). On efficient equilibrium solutions in dynamic games of resource management. *Resources and Energy,* 11, 23–34.

Dockner, E and NV Long (1993). International pollution control: Cooperative versus non-cooperative strategies. *Journal of Environmental Economics and Management,* 25, 13–29.

Dockner, E, NV Long and Sorger (1996). Analysis of Nash equilibria in a class of capital accumulation games. *Journal of Economic Dynamics and Control,* 20, 1209–35.

Dockner, E and G Mosburger (2007). Capital accumulation, asset values and imperfect product market competition. *Journal of Difference Equations and Applications,* 13(2), 197–215.

Dockner, E and K Nishimura (1999). Transboundary Pollution in a Dynamic Game Model. *Japanese Economic Review,* 50, 443–456.

Dockner, E and G Sorger (1996). Existence and properties of equilibria for a dynamic game on productive assets. *Journal of Economic Theory,* 71, 209–227.

Dornbusch, R (1976). Expectations and exchange rate dynamics. *Journal of Political Economy,* 84, 1161–1176.

Dornbusch, R (1987). Exchange rates and prices. *American Economic Review,* 77(1), 93–106.

Driffill, J (1982). Optimal money and exchange rate policies. *Greek Economic Review,* 4, 261–283.

Driskill, R and A Horowitz (1996). Durability and strategic trade: Are there rents to be captured? *Journal of International Economics,* 41(1), 179–194.

Driskill, R and S McCafferty (1989a). Dynamic duopoly with adjustment costs: A differential game approach. *Journal of Economic Theory,* 69, 324–338.

Driskill, R and S McCafferty (1989b). Dynamic duopoly with output adjustment costs in international markets: Taking the conjecture out of conjectural variations. In *Trade Policies for International Competitiveness,* R Feenstra (ed.), pp. 124–143. University of Chicago: Chicago Press.

Driskill, R and S McCafferty (1996). Industrial policy and duopolistic trade with dynamic demand. *Review of Industrial Organization,* 11, 355–373.

Driskill, R and S McCafferty (2001). Monopoly and oligopoly provision of addictive goods. *International Economic Review,* 42(1), 43–72.

Duffie, DY Geanakopolos, A MasColell and A McLennan (1994). Stationary Markov equilibrium. *Econometrica,* 62, 745–781.

Dutta, PK, S Lach and A Rustichini (1995). Better late than early: Vertical differentiation in the adoption of new technology. *Journal of Economics & Management Strategies,* 4(4), 563–598.

Dutta, PK and R Radner (2006). Population growth and technological change in a global warming model. *Economic Theory,* 29, 251–270.

Dutta, PK and R Radner (2009). Strategic analysis of global warming: Theory and some numbers. *Journal of Economic Behavior and Organization,* forthcoming.

Dutta, PK and R Sundaram (1992). Markovian equilibrium in a class of stochastic games: Existence theorems for discounted and undiscounted models. *Economic Theory*, 2, 197–214.

Dutta, PK and R Sundaram (1993a). The tragedies of the commons. *Economic Theory*, 3, 413–426.

Dutta, PK and R Sundaram (1993b). How different can strategic models be? *Journal of Economic Theory*, 60, 42–61.

Eaton, J and GM Grossman (1985). Tariffs as insurance: Optimal commercial policy when domestic markets are incomplete. *Canadian Journal of Economics*, 18, 258–272.

Eaton, J and GM Grossman (1986). Optimal trade and industrial policy under oligopoly. *Quarterly Journal of Economics*, 101, 383–406.

Eaton, J and GM Grossman (1988). Trade and industrial policy under oligopoly: Reply. *Quarterly Journal of Economics*, 103, 603–607.

Ehtamo, H and RP Hämäläinen (1989). Incentive strategies and equilibria for dynamic games with delayed information. *Journal of Optimization Theory and Application*, 63, 355–369.

Ehtamo, H and RP Hämäläinen (1993). A cooperative incentive equilibrium for a resource management problem. *Journal of Economic Dynamics and Control*, 17, 659–678.

Ellis, C, O Dincer and G Waddel (2005). Corruption, Decentralization and Yardstick Competition. University of Oregon Working Paper.

Eswaran, M and T Lewis (1985). Exhaustible resources and alternative equilibrium concepts. *Canadian Journal of Economics*, 18, 459–473.

Eyckman, J (2001). On the Far-Sighted Stability of the Kyoto Protocal. *CLIMNEG* Working Paper 40, CORE, Université Catholique de Louvain.

Ferejohn, J (1986). Incumbent Performance and Electoral Control. *Public Choice*, 50, 5–26.

Fershtman, C and E Muller (1984). Capital accumulation games of infinite duration. *Journal of Economic Theory*, 33, 322–339.

Fershtman, C and M Kamien (1987). Dynamic duopolistic competition with sticky prices. *Econometrica*, 55, 1151–1164.

Fershtman, C and S Nitzan (1991). Dynamic voluntary provision of public goods. *European Economic Review*, 35, 1057–1067

Feichtinger, G (1982). Optimal policies for two firms in a non-cooperative research project. In *Optimal Control Theory and Economic Analysis*, G Feichtinger (ed.), pp. 373–393. Amsterdam: North Holland.

Feichtinger, G and F Wirl (1993). A dynamic variant of the battle of the sexes. *International Journal of Game Theory*, 22, 359–380.

Ferris, M and T Munson (2000). Complementary problems in GAMS and the PATH solver. *Journal of Economic Dynamics and Control*, 24, 165–188.

Figuières, C (2009). Markov interactions in a class of dynamic games. *Theory and Decision*, 66(1), 39–68.

Figuières, C, NV Long and M Tidball (2009). Sustainability and the mixed Bentham-Rawls criterion: A discrete time characterization, Typescript, McGill University.

Finus, M (2001). *Game Theory and International Environmental Cooperation.* Cheltenham, UK: Edward Elgar.

Fischer, RD and L Mirman (1986). The complete fish wars: Biological and dynamic interactions. *Journal of Environmental Economics and Management*, 30, 34–42.

Fischer, R and L Mirman (1992). Strategic dynamic interaction: Fish wars. *Journal of Economic Dynamics and Control*, 16, 267–287.

Flaherty, MT (1980a). Dynamic limit pricing, barriers to entry, and rational firms. *Journal of Economic Theory*, 23, 160–182.

Flaherty, MT (1980b). Industry structure and cost-reducing investment. *Econometrica*, 48, 1187–1209.

Food and Agriculture Organization (2009). *The State of World Fisheries and Aquaculture 2008*. FAO Fisheries and Aquaculture Department Food and Agriculture Organization of the United Nations.

Forster, BA (1972). A note on optimal control of pollution. *Journal of Economc Theory*, 5(3), 537–539.

Forster, BA (1973). Optimal capital accumulation in a polluted environment. *Southern Economic Journal*, 39, 544–547.

Fudenberg, D and DK Levine (1988). Open-loop and closed-loop equilibria in dynamic games with many players. *Journal of Economic Theory*, 44, 1–18.

Fudenberg, D and J Tirole (1983). Learning by doing and market performance. *Bell Journal of Economics*, 14, 522–530.

Fudenberg, D and J Tirole (1985). Preemption and rent equalization in the adoption of new technology. *Review of Economic Studies*, 52(3), 383–401.

Fudenberg, D and J Tirole (2000). *Game Theory*. Cambridge, MA: MIT Press.

Fujiwara, K (2009a). Gains from trade in a differential game model of asymmetric oligopoly. *Review of International Economics*, 17(5), 1066–1073.

Fujiwara, K (2009b). A dynamic reciprocal dumping model of international trade. *Asia-Pacific Journal of Accounting and Economics*, 16(3), 255–270.

Fujiwara, K (2010). Losses from competition in a dynamic game model of renewable resource oligopoly. *Resources and Energy Economics*, forthcoming.

Fujiwara, K and N Matsueda (2010). Effects of transboundary stock pollution on the mode of international competition. *Asia-Pacific Journal of Accounting and Economics*, 17(3).

Fujiwara, K and NV Long (2009a). Feedback Stackelberg equilibria in an exhaustible resource market: The price-setting case, Working Paper, McGill University.

Fujiwara, K and NV Long (2009b). Feedback Stackelberg equilibria in an exhaustible resource market: The quantity-setting case, Working Paper, McGill University.

Fujiwara, K and NV Long (2009c). Pigouvian taxes and rent shifting: comparing Nash equilibrium, global feedback Stackelberg equilibria and stagewise Stackelberg equilibria, Working Paper, McGill University.

Fujiwara, K and NV Long (2010a). Welfare effects of reducing home bias in government procurements: A dynamic contest model. *Review of Development Economics*, forthcoming.

Fujiwara, K and NV Long (2010b). Welfare implications of leadership in a resource market under bilateral monopoly. Cirano Working Paper 2010s–16, Cirano, Montrial.

Gallini, N, T Lewis and R Ware (1983). Strategic timing and pricing of a substitute in a cartelized resource market. *Canadian Journal of Economics*, 16, 429–446.

Galor, O (1996). Convergences? Inferences from theoretical models. *Economic Journal*, 106, 1056–1069.

Gaudet, G and P Lasserre (2009). The efficient use of multiple sources of a non-renewable resource under supply cost uncertainty. *International Economic Review*, forthcoming.

Gaudet, G, P Lasserre and NV Long (1995). Optimal resource royalties with unknown and temporally independent extraction cost structure. *International Economic Review*, 36, 715–749.

Gaudet, G, P Lasserre and NV Long (1996). Dynamic incentive contracts with uncorrelated private information and history-dependent outcomes. *The Japanese Economic Review*, 47, 321–334.

Gaudet, G, P Lasserre and NV Long (1998). Real investment decisions under adjustment costs and asymmetric information. *Journal of Economic Dynamics and Control*, 23, 71–95.

Gaudet, G and H Lohoues (2008). On limits to the use of linear Markov strategies in common property natural resource games. *Environmental Modelling and Assessment*, 13, 567–574.

Gaudet, G and NV Long (1994). On the effects of the distribution of initial endowments in a non-renewable resource duopoly. *Journal of Economic Dynamics and Control*, 18, 1189–1198.

Gaudet, G and NV Long (2003). Recycling redux: A Nash-Cournot approach. *Japanese Economic Review*, 54, 409–419.

Gaudet, G, M Moreaux and S Salant (2001). Intertemporal depletion of resource sites by spatially distributed users. *American Economic Review*, 91, 1149–1159.

Germain, MP Toint, H Tulkens and A de Zeeuw (2003). Transfers to sustain core-theoretic cooperation in international stock-pollutant control. *Journal of Economic Dynamics and Control*, 28, 79–99.

Gilbert, R (1978). Dominant firm pricing policy in a market for an exhaustible resource. *The Bell Journal of Economics*, 9(2), 385–395.

Gilbert, R and S Goldman (1978). Potential competition and the monopoly price of an exhaustible resource. *Journal of Economic Theory*, 17, 319–331.

Goldsmith, OS (1974). Market allocation of exhaustible resources. *Journal of Political Economy*, 82(5), 1035–1040.

Gollier, C and R Zeckhauser (2005). Aggregation of Heterogeneous Time Preferences. *Journal of Political Economy*, 113, 878–896.

Gordon, HS (1954). Economic theory of common property resources. *Journal of Political Economy*, 62(2), 124–142.

Gradstein, M (2007). Inequality, democracy and the protection of property rights. *Economic Journal*, 117, 252–269.

Grafton, Q, T Kompas and R Hilborn (2007). The Economics of Over-exploitation Revisited. *Science*, 7 December 2007, p. 1601.

Green, E and RA Porter (1984). Non-cooperatice collusion under imperfect price information. *Econometrica*, 52, 87–100.

Greenberg, J (1986). Stable Standards of Behavior: A Unifying Approach to Solution Concepts, Technical Report No. 484, Institute for Mathematical Studies in the Social Sciences, Standford University.

Greenberg, J (1990). *The Theory of Social Situations: An Alternative Game-Theoretic Approach*. New York: Cambridge University Press.

Groot, F, C Withagen and A de Zeeuw (1992). Note on the open-loop von Stackelberg equilibrium in the cartel-versus-fringe model. *Economic Journal*, 102, 1478–1484.

Groot, F, C Withagen and A de Zeeuw (2003). Strong time-consistency in the cartel-versus-fringe Model. *Journal of Economic Dynamics and Control*, 28, 287–306.

Gruver, G (1976). Optimal investment and pollution control in a neoclassical growth context. *Journal of Environmental Economics and Management*, 5, 165–177.

Hämäläinen, RP, A Haurie and V Kaitala (1985). Equilibria and Threats in a Fishery Management Game. *Optimal Control Applications and Methods*, 6, 315–333.

Hämäläinen, RP, J Ruusunen and V Kaitala (1986). Myopic Stackelberg equilibria and social coordination in share contract fishery. *Marine Resource Economics*, 3, 209–235.

Hämäläinen, RP, J Ruusunen and V Kaitala (1990). Cartel and dynamic contract in sharefishing. *Journal of Environmental Economics and Management*, 19, 175–192.

Hardin, G (1968). The tragedy of the commons. *Science*, 162, 1243–48.

Harris, R (1985). Why voluntary export restraints are 'voluntary'? *Canadian Journal of Economics*, 18, 799–809.

Harris, C and J Vickers (1995). Innovation and natural resources: A dynamic game with uncertainty. *Rand Journal of Economics*, 26, 418–430.

Hartwick, JM (1977). Intergenerational equity and the investing of rents from exhaustible resources. *American Economic Review*, 69, 972–974.

Haurie, A (1995). Environmental coordination in dynamic oligopolistic markets. *Group Decision and Negotiation*, 4, 46–67.

Haurie, A and JB Krawczyk (1997). Optimal charges on river effluent from lump and distributed sources. *Environmental Modeling and Assessment*, 2, 177–199.

Haurie, A, JB Krawczyk and M Roche (1994). Monitoring cooperative equilibria in a stochastic differential game. *Journal of Optimization Theory and Applications*, 81, 73–95.

Haurie, A, F Moresino and L Viguier (2006). A two-level differential game of international emssions trading. In *Advances in Dynamic Games, Annals of the International Society for Dynamic Games*, A Haurie, S Muto, LA Petrosjan and TES Raghavan (eds.), 8, 293–307.

Haurie, A and M Pohjola (1987). Efficient equilibria in a differential game of capitalism. *Journal of Economic Dynamics and Control*, 11, 65–78.

Haurie, A and B Towinski (1990). Cooperative equilibria in discounted stochastic differential games. *Journal of Optimization Theory and Applications*, 64(3).

Haurie, A and L Viguier (2003). A stochastic dynamic game of carbon emissions trading. *Environmental Modeling and Assessment*, 8, 293–248.

Haurie, A and G Zaccour (1995). Differential game models of global environmental management. *Annals of the International Society of Dynamic Games*, Vol. 2, pp. 3–24.

Helm, C (2003). International emissions trading with endogenous allowance choices. *Journal of Public Economics*, 87, 2737–2747.

Hilborn, R, JM Orensanz and AM Parma (2005). Institutions, incentives and the future of fisheries. *Philosophical Transactions of the Royal Society* B, 360, 47.

Hiller, B and JM Malcomson (1984). Dynamic inconsistency, rational expectations and optimal government policies. *Econometrica*, 52, 1437–1453.

Hillman, AL and NV Long (1983). Pricing and depletion of an exhaustible resource when there is anticipation of trade disruption. *Quarterly Journal of Economics*, 98(2), 215–233.

Hillman, AL and NV Long (1985). Monopolistic recycling of oil revenue and intertemporal bias in oil depletion and trade. *Quarterly Journal of Economics*, 100(3), 597–624

Hirshleifer, J (1983). From weakest link to best-shot: The voluntary provision of public goods. *Public Choice*, 41, 371–386.

Hoel, M (1978a). *Resource Extraction under Some Alternative Market Structures.* Mathematical Systems in Economics, Vol. 39, Meisenheim an Glan: Verlag Anton Hain.

Hoel, M (1978b). Distribution and growth as a differential game between workers and capitalists. *International Economic Review*, 9, 335–350.

Hoel, M (1979). *Extraction of an Exhaustible Resource under Uncertainty.* Mathematical Systems in Economics, Vol. 51, Meisenheim an Glan: Verlag Anton Hain.

Hoel, M. (1991). Global environmental problems: The effects of unilateral actions taken by one country. *Journal of Environmental Economics and Management*, 20(1), 55–70.

Hoel, M. (1992). Emission taxes in a dynamic international game of CO_2 emissions. In *Conflicts and Cooperations in Managing Environmental Resources*, R Pethig (ed.), Berlin: Springer-Verlag.

Hoel, M (1993). Intertemporal properties of an international carbon tax. *Resource and Energy Economics*, 15(1), 51–70.

Hoppe, HC and U Lehman-Grube (2001). Second-mover advantages in dynamic quality competition. *Journal of Economics & Management Strategies*, 10, 419–433.

Hoppe, HC and U Lehman-Grube (2005). Innovation timing games: A general framework with applications. *Journal of Economic Theory*, 121, 30–50.

Hughes Hallet, AJ (1984). Non-cooperative strategies for dynamic policy games and the problem of time inconsistency. *Oxford Economic Papers*, 56, 381–399.

Huizinga, H and SB Nielsen (1997). Capital income and profit taxation with foreign ownership of firms. *Journal of International Economics*, 42, 149–165.

Hung, NM, MC Kemp and NV Long (1984). On the transition from an exhaustible resource stock to an inexhaustible substitute Essay 6. In *Essays in the Economics of Exhaustible Resources*, MC Kemp and NV Long (eds.), pp. 104–143. Amsterdam: North Holland.

Hung, NM and NV Quyen (1990). Bilateral Monopoly Games and Endogenous Leadership: The Case of Innovation in Resource Economics, Cahier 9101, GREEN, Université Laval.

Hung, NM and NV Quyen (1993). On R&D timing under uncertainty: The case of exhaustible resource substitution. *Journal of Economic Dynamics and Control*, 17(5&6), 971–991.

Hung, NM and NV Quyen (1994). *Dynamic Timing Decisions under Uncertainty.* Lecture Notes in Economics and Mathematical Systems. New York: Springer.

Hwang, H and C-C Mai (1988). On the equivalence of tariff and quota under duopoly. *Journal of International Economics*, 24, 373–380.

Isaacs, R (1965). *Differential Games.* New York: John Wiley and Sons.

Issacs, R (1969). Differential games: Their scope, nature, and future. *Journal of Optimization Theory and Applications*, 3(5), 283–295.

Itaya, J and D Dasgupta (1995). Dynamics, consistent conjectures, and heterogeneous agents in the private provision of public goods. *Public Finance*, 50, 371–389.

Itaya, J and K Shimomura (2001). A dynamic conjectural variations model in the private provision of public goods: A differential game approach. *Journal of Public Economics*, 81, 153–172.

Jenson, R and EF Toma (1991). Debt in a model of tax competition. *Regional Science and Urban Economics*, 21, 371–392.

Johnson, HG (1954). Optimal tariffs and retaliation. *Review of Economic Studies*, 21, 142–153.

Johnson, HG (1972). *Aspects of the Theory of Tariffs.* Cambridge, MA: Harvard University Press.

Jørgensen, S and G Sorger (1990). Feedback Nash equilibria in a problem of optimal fishery management. *Journal of Optimization Theory and Applications*, 64, 293–310.

Jørgensen, S and DWK Yeung (1996). Stochastic differential game model of common property fishery. *Journal of Optimization Theory and Applications*, 90(2), 381–403.

Jørgensen, S and G Zaccour (2001a). Time-consistent side payments in a dynamic game of downstream pollution. *Journal of Economic Dynamics and Control*, 25(12), 1973–1987.

Jørgensen, S and G Zaccour (2001b). Incentive equilibrium strategies and welfare allocation in a dynamic game of pollution control. *Automatica*, 37(1), 29–36.

Jørgensen, S and G Zaccour (2004). *Differential Games in Marketing*. Boston: Kluwer Academic Publishers.

Jørgensen, S and G Zaccour (2007). Development in differential game theory and numerical methods: Economic and management applications. *Computational Management Science*, 4(2), 159–181.

Judd, KL (1985). Redistributive taxation in a simple perfect foresight model. *Journal of Public Economics*, 28, 59–83.

Judd, KL (1992). Projection methods for solving aggregate growth models. *Journal of Economic Theory*, 58, 410–452.

Judd, KL (1997). Computational economics and economic theory: Substitutes or complements? *Journal of Economic Dynamics and Control*, 21, 907–942.

Judd, KL (1998). *Numerical Methods in Economics*. Cambridge, MA: MIT Press.

Jun, B and X Vives (2004). Strategic incentives in dynamic duopoly. *Journal of Economic Theory*, 116, 249–281

Kahn, C (1986). The durable goods monopolist and consistency with increasing cost. *Econometrica*, 54, 274–294.

Kaitala, V (1986). Game theory model of fishery management — A survey. In *Dynamic Games and Applications in Economics*, T Başar (ed.), pp. 252–256. Berlin and New York: Springer.

Kaitala, V (1989). Non-uniqueness of no-memory feedback Nash equilibria in a fishery resource game. *Automatica*, 25, 587–592.

Kaitala, V (1993). Equilibria in a stochastic resource management game under imperfect information. *European Journal of Operational Research*, 71, 439–453.

Kaitala, V, KG Mäulet and H Tulkens (1995). The acid rain game as a resource allocation process with an application to the international cooperation among Finland, Russia, and Estonia. *Scandinavian Journal of Economics*, 97, 325–343.

Kaitala, V and M Pohjola (1995). Sustainable international agreements on greenhouse warming: A game theory study. *Annals of the International Society of Dynamic Games*, 2, 67–87.

Kaitala, V, M Pohjola and O Tahvonen (1991). Transboundary air pollution between Findland and the USSR: A dynamic acid rain game. In *Dynamic Games in Economic Analysis*, R Hämäläinen and H Ehtamo (eds.). pp. 183–192. Berlin: Springer.

Kaitala, V, M Pohjola and O Tahvonen (1992a). Transboundary air pollution and soil acidification: A dynamic analysis of an acid rain game between Finland and the USSR. *Environmental and Resource Economics*, 2(2), 161–181.

Kamien, MI and NL Schwartz (1991). *Dynamic Optimization*. Amsterdam: North Holland.

Kaitala, V, M Pohjola and O Tahvonen (1992b). An economic analysis of transboundary air pollution between Finland and the Soviet Union. *Scandinavian Journal of Economics*, 94, 409–424.

Karp, LS (1984). Optimality and consistency in a differential game with non-renewable resources. *Journal of Economic Dynamics and Control*, 8, 73–97.

Karp, LS (1987). Consistent tariffs with dynamic supply responses. *Journal of International Economics*, 23, 369–376.

Karp, LS (1993). Open-loop and feedback models of dynamic oligopoly. *International Journal of Industrial Organization*, 11, 369–389.

Karp, LS (1996). Monopoly power can be disadvantageous in the extraction of a durable non-renewable resource. *International Economic Review*, 37(4), 925–949.

Karp, LS (2005). Nonpoint source pollution taxes and excessive tax burden. *Environmental & Resource Economics*, 31, 229–251.

Karp, L and IH Lee (2003). Time-consistent policies. *Journal of Economic Theory*, 112, 353–364.

Karp, LS and J Livernois (1992). On efficiency-inducing taxation for a non-renewable resource monopolist. *Journal of Public Economics*, 49, 219–239.

Karp, LS and J Livernois (1994). Using automatic tax changes to control pollution emissions. *Journal of Environmental Economics and Management*, 27, 38–48.

Karp, LS and S Sachetti (1997). Dynamics and limited cooperation in international environmental agreements. Mimeo, University of California, Berkeley.

Karp, L and DM Newbery (1991). Optimal tariffs on exhaustible resources. *Journal of International Economics*, 30, 285–299.

Karp, L and DM Newbery (1992). Dynamically consistent oil import tariffs. *Canadian Journal of Economics*, 25, 1–21.

Katayama, S and NV Long (2010). A dynamic game of status-seeking with public capital and an exhaustible resource. *Optimal Control, Applications and Methods*, 31, 43–53.

Katz, ML and C Shapiro (1987). R&D Rivalry with licensing or imitation. *American Economic Review*, 77(3), 402–420.

Keeler, E, M Spence and R Zeckhauser (1971). The optimal control of pollution. *Journal of Economic Theory*, 4, 19–34.

Kemp MC (1976). How to eat a cake of unknown size (Chapter 23). In *Three Topics in the Theory of International Trade*, MC Kemp (ed.). Amsterdam: North Holland.

Kemp, MC (1977). Further generalizations of the cake-eating problem under uncertainty. *Theory and Decision*, 8, 363–367.

Kemp, MC and NV Long (1978). Optimal consumption of depletable resources: Comments. *Quarterly Journal of Economics*, 92, 345–353.

Kemp, MC and NV Long (1979). The interaction between resource-poor and resource-rich economies. *Australian Economic Papers*, 18, 258–267, reprinted as Essay 17. In *Exhaustible Resources, Optimality and Trade*,

1980, MC Kemp and NV Long (eds.), pp. 197–209. Amsterdam: North Holand.

Kemp, MC and NV Long (1980a). *Exhaustible Resources, Optimality and Trade.* Amsterdam: North Holland.

Kemp, MC and NV Long (1980b). Eating a cake of unknown size: Pure competition versus social planning, Essay 5. In *Exhaustible Resources, Optimality, and Trade*, MC Kemp and NV Long (eds.), pp. 55–70. Amsterdam: North Holland.

Kemp, MC and NV Long (1980c). Resource extraction under condition of common access, Essay 10. In *Exhaustible Resources, Optimality, and Trade*, MC Kemp and NV Long (eds.). Amsterdam: North Holland.

Kemp, MC and NV Long (1980d). Optimal tariff and exhaustible resources, Essay 16. In *Exhaustible Resources, Optimality and Trade*, MC Kemp and NV Long (eds.). Amsterdam: North Holland.

Kemp, MC and NV Long (1984a). *Essays in the Economics of Exhaustible Resources.* Amsterdam: North Holland.

Kemp, MC and NV Long (1984b). The role of natural resources in trade models. In *Handbook of International Economics*, Vol. 1, RW Jones and P Kenen (eds.), pp. 367–417. New York: Elsevier Science, North-Holland.

Kemp, MC and NV Long (1985). Eating a cake of unknown size: The case of a growing cake. *Economic Letters*, 18(2), 67–69.

Kemp, MC and NV Long (2009). Foreign aid in the presence of corruption: A differential game. *Review of International Economics*, 17(2), 230–243.

Kemp, MC, NV Long and K Shimomura (1991). *Labour unions and the theory of international trade.* Amsterdam: North Holland.

Kemp, MC, NV Long and K Shimomura (1992). A dynamic formulation of the foreign aid process. In *Dynamic Economic Models and Optimal Control*, G Feichtinger, (ed.). Amsterdam: North-Holland.

Kemp, MC, NV Long and K Shimomura (1993). Cyclical and non-cyclical redistributive taxation. *International Economic Review*, 34, 415–429.

Kemp, MC, NV Long and K Shimomura (2001). A differential model of tariff war. *Japan and the World Economy*, 13(3), 279–298.

Kennedy, P (1987). *The Rise and Fall of the Great Powers.* New York: Random House.

Kessing, S (2007). Strategic complimentarity in the dynamic private provision of a discrete public good. *Journal of Public Economic Theory*, 9, 699–710.

Keutiben, O (2009). On Capturing Foreign Oil Rents, Typescript, Université de Montréal.

Khalatbari, F (1977). Market imperfection and optimal rate of depletion of natural resources. *Economica*, 44, 409–418.

Köthenbürger, M and B Lockwood (2007). Does tax competition really promote growth? CESifo Working Paper 2102, University of Munich.

Koulovatianos, C (2007). Rational Noncooperative Strategic Exploitation of Species in a Predator-Prey Ecosystem with Random Disturbances, Typescript, University of Vienna.

Koulovatianos, C and L Mirman (2007). The effects of market structure on industry growth: Rivalrous non-excludable capital. *Journal of Economic Theory*, 133(1), 199–218.

Kompas, T, N Che and Q Grafton (2008). Fisheries instrument choice under uncertainty. *Land Economics*, 84, 652–666.

Krawczyk, JB (2005). Coupled constraint Nash equilibria in environmental games. *Resource and Energy Economics*, 27, 157–181.

Krawczyk, JB and K Shimomura (2003). Why countries with the same technology and preferences can have different growth rates? *Journal of Economics Dynamics and Control*, 27, 1899–1916.

Krawczyk, JB and B Tolwinski (1993). A cooperative solution for the three nation problem of exploitation in the southern bluefin tuna. *IMA Journal of Mathematics Applied in Medecine and Biology*, 10, 135–147.

Kreps, DM (1990). Corporate culture and economic theory. In *Perspective on Positive Political Economy*, J Alt and K Shepsle (eds.). Cambridge, UK: Cambridge University Press.

Kurzban, R and D Houser (2004). Experiments investigating cooperative types in humans: A complement to evolutionary theory and simulations. *Proceedings of the National Academy of Sciences*, 102(5), 1803–1807.

Kydland, F (1975). Noncooperative and dominant player solutions in discrete dynamic games. *International Economic Review*, 16, 321–335.

Kydland, F and EC Prescott (1977). Rules rather than discretion: The inconsistency of optimal plans. *Journal of Political Economy*, 85, 473–491.

Lahiri, S and Y Ono (1988). Helping minor firms reduce welfare. *Economic Journal*, 98, 1199–1202.

Lambertini, L and G Rossini (1998). A dynamic model of differentiated oligopoly with capital accumulation. *Journal of Economic Theory*, 83(1), 145–155.

Lancaster, K (1973). The dynamic inefficiency of capitalism. *Journal of Political Economy*, 81, 1092–1109.

Lane, J and W Leininger (1986). Price characterization and Pareto efficiency of game-equilibrium growth. *Zeitschrift für Nationalökonomie*, 46, 34.

Lane, J and T Mitra (1981). On Nash equilibrium programmes of capital accumulation under altruistic preferences. *International Economic Review*, 22, 309–331.

Lane, P and A Tornell (1996). Power, growth, and the voracity effect. *Journal of Economic Growth*, 1, 213–241.

Lansing, KJ (1999). Optimal redistributive taxation in a neoclassical growth model. *Journal of Public Economics*, 73, 423–453.

Lapan, H (1988). The optimal tariff, production lags, and time consistency. *American Economic Review*, 78(3), 395–401.

Laussel, D, M de Montmarin and NV Long (2004). Dynamic duopoly with congestion effects. *International Journal of Industrial Organization*, 22(5), 655–677.

Leahy, D and JP Neary (1999). Learning by doing, precommitment and infant industry promotion. *Review of Economic Studies*, 66, 447–474.

Ledyard, J (1995). Public goods: A survey of experimental research. In *Handbook of Experimental Game Theory*, J Kagel and A Roth (eds.). Princeton: Princeton University Press.

Lee, K (1997). Tax competition with imperfectly mobile capital. *Journal of Urban Economics*, 42, 222–242.

Lee, T and LL Wilde (1980). Market structure and innovation: A reformulation. *Quarterly Journal of Economics*, 94(2), 429–436.

Leininger, W (1985). Rawls' maximin criteria and time-consistency: Further results. *Review of Economic Studies*, 52, 505–513.

Leininger, W (1986). The existence of perfect equilibria in a model of growth with altruism between generations. *Review of Economic Studies*, 53, 349–67.

Léonard, D and NV Long (1992). *Optimal Control Theory and Static Maximization in Economics*. Cambridge, UK: Cambridge University Press.

Léonard, D and NV Long (2009). Endogenous Changes in Property Rights Regime, Typecript, Flinders University.

Leour, A and H Verbon (1997). Tax competition and redistribution in a two-country endogenous growth model. *International Tax and Public Finance*, 4, 485–497.

Levhari, D and L Mirman (1980). The great fish wars: An example of using a dynamic Nash-Cournot solution. *Bell Journal of Economics*, 11, 322–34.

Lewis, T and R Schmalensee (1980). On oilgopolistic markets for non-renewable natural resources. *Quarterly Journal of Economics*, 95, 475–491.

Liski, M and O Tahvonen (2004). Can carbon tax eat OPEC's rents? *Journal of Environmental Economics and Management*, 47, 1–12.

List, JA and CF Mason (1999). Spatial aspects of pollution control when pollutants have synergistic effects: Evidence from a differential game with asymmetric information. *Annals of Regional Science*, 33(4), 439–452.

List, JA and CF Mason (2001). Optimal institutional arrangements for transboundary pollutants in a second-best world: Evidence from a differential game with asymmetric players. *Journal of Environmental Economics and Management*, 42(3), 277–296.

Lockwood, B (1996). Uniqueness of Markov-perfect equilibrium in infinite-time affine-quadratic differential Games. *Journal of Economic Dynamics and Control*, 20, 751–765.

Loeb, M and WA Magat (1979). A decentralized method for utility regulation. *Journal of Law and Economics*, 22, 399–404.

Long, NV (1975). Resource extraction under the uncertainty about possible nationalization. *Journal of Economic Theory*, 10(1), 42–53.

Long, NV (1994). On optimal enclosure and optimal timing of enclosure. *Economic Record*, 70(221), 141–145.

Long, NV (1992). Pollution control: A differential game approach. *Annals of Operations Research*, 37, 283–296.

Long, NV (2006a). Capacity utilization and investment in environmental quality. *Environmental Modeling and Assessment*, 11(2), 166–177.

Long, NV (2006b). Sustainable development with intergenerational and international cooperation. In *Economic Development, Climate Change, and the Environment.* A Sinha and S Mitra (eds.), pp. 116–148. London: Routledge.

Long, NV (2007). Toward a theory of a just savings principle. In *Intergenerational Equity and Sustainability*, J Roemer and K Suzumura (eds.), pp. 291–319. London: Palgrave.

Long, NV and S Katayama (2002). Common property resource and private capital accumulation. In *Optimal Control and Differential Games*, G Zaccour (ed.) pp. 193–209. London: Kluwer Academic Publishers.

Long, NV and S McWhinnie (2010). The tragedy of the commons when relative performance matters. Working Paper 2010–2007, University of Adelaide.

Long, NV and B Sengupta (2008). Electoral incentives, institutions and rent-seeking. *Indian Growth and Development Review*, 1(2), 133–146.

Long, NV and K Shimomura (1998). Some results on the Markov equilibria of a class of homogeneous differential games. *Journal of Economic Behavior and Organization*, 33(3), 557–566.

Long, NV and K Shimomura (2004a). Relative wealth, status seeking, and catching up. *Journal of Economic Behavior and Organization*, 53(4), 529–542.

Long, NV and K Shimomura (2004b). Relative wealth, catching up, and economic growth. In *Economic Growth and Macroeconomic Dynamics*, S Dowrick, R Pitchford and S Turnovsky (eds.), pp. 18–45. Cambridge, UK: Cambridge University Press.

Long, NV and K Shimomura (2000). Semi-stationary equilibria in leader-follower games, CIRANO Working Paper 2000s-08.

Long, NV and K Shimomura (2002). Redistributive taxation in closed and open economies, Chapter 7. In *Economic Theory and International Trade: Essays in Honour of Murray C. Kemp*, A Woodland (ed.) pp. 104–123. Cheltenham, UK: Edward Elgard.

Long, NV, K Shimomura, and H Takahashi (1999). Comparing open loop and Markov perfect equilibria in a class of differential games. *Japanese Economic Review*, 50(4), 457–469.

Long, NV and H-W Sinn (1985). Surprise price shift, tax changes and the supply behaviour of resource extracting firms. *Australian Economic Papers*, 24(45), 278–289.

Long, NV and G Sorger (2006). Insecure property rights and growth: The roles of appropriation costs, wealth effects, and heterogeneity. *Economic Theory*, 28(3), 513–529.

Long, NV and G Sorger (2009). A dynamic principal-agent problem as a feedback Stackelberg differential game, Working Paper 0905, Department of Economics, University of Vienna.

Long, NV and A Soubeyran (2001). Cost manipulation games in oligopoly, with cost of manipulating. *International Economic Review*, 42(2), 505–533.

Long, NV and NJ Vousden (1977). Optimal control theorems, Essay 1. In *Applications of Control Theory to Economic Analysis*, JD Pitchford and SJ Turnovsky (eds.), pp. 11–34. Amsterdam: North-Holland.

Long, NV and S Wang (2009). Resource-grabbing by status-conscious Agents. *Journal of Development Economics*, 89(1), 39–50.

Long, NV and K-Y Wong (1997). Endogenous growth and international trade: A survey. In *Dynamics, Trade, and Growth*, B Jensen and K-Y Wong (eds.), pp. 11–74. Ann Arbor, Michigan: University of Michigan Press.

Loschel, A and Z Zhang (2002). The Economic and Environmental Implications of the US Repudiation of the Kyoto Protocol and the Subsequent Deals in Bonn and Marrakech, Nota Di Lavoro 23.2002, Fondazione Enie Enrico Mattei (FEEM).

Loury, G (1986). A theory of oiligopoly: Cournot equilibrium in exhaustible resource markets with fixed supplies. *International Economic Review*, 27, 285–301.

Lucas, RE and N Stokey (1984). Optimal growth with many consumers. *Journal of Economic Theory*, 32, 139–171.

Maddison, A (1982). *Phases of Capitalist Development.* New York: Oxford University Press.

Mai, C-C and H Hwang (1988). Why voluntary export restraints are voluntary: An extension. *Canadian Journal of Economics*, 21, 877–882.

Majumdar, KM and RK Sundaram (1991). Symmetric stochastic games of resource extraction: The existence of non-randomized stationary equilibrium. In *Stochastic Games and Related Topics*, TES Raghavan (ed.), pp. 175–190. The Netherlands: Kluwer Academic.

Maler, K-G and A de Zeeuw (1998). The acid rain differential game. *Environmental and Resource Economics*, 12, 167–184.

Malueg, DA and SO Tsutsui (1997). Dynamic R&D competition with learning. *Rand Journal of Economics*, 28(4), 751–772.

Martin, WE, RH Patrick and B Tolwinski (1993). A dynamic game of a transboundary pollutant with asymmetric players. *Journal of Environmental Economics and Management*, 24, 1–12.

Martín-Herrán, G and J Rincon-Zapareto (2005). Efficient Markov-perfect Nash equilibria: Theory and application to dynamic fishery games. *Journal of Economic Dynamics and Control*, 29, 1073–1096.

Marx, LM and SA Matthews (2000). Dynamic voluntary contribution to a public project. *Review of Economic Studies*, 67, 327–358.

Maskin, E and D Newbery (1990). Disadvantageous oil tariffs and dynamic consistency. *American Economic Review*, 80(1), 143–156.

Maskin, E and J Tirole (1987). A theory of dynamic oligopoly: III: Cournot competition. *European Economic Review*, 31, 947–968.

Maskin, E and J Tirole (1988a). A theory of dynamic oligopoly. I: overview and quantity competition with large fixed costs. *Econometrica*, 56, 549–569.

Maskin, E and J Tirole (1988b). A theory of dynamic oligopoly, II: Price competition, kinked demand curves and Edgeworth cycles. *Econometrica*, 56, 571–599.

Maskin, E and J Tirole (2001). Markov-perfect equilibrium I: Observable actions. *Journal of Economic Theory*, 100, 191–219.

Matsuyama, K (1990). Perfect equilibria in a trade liberalization game. *American Economic Review*, 80, 480–492.

McLenaghan, R and S Levy (1996). Geometry. In *CRC Standard Mathematical Tables and Formulae*, D Zwillinger (Ed.). Boca Raton, Florida: CRC Press.

McWhinnie, SF (2009). The tragedy of the commons in international fisheries: An empirical investigation. *Journal of Environmental Economics and Management*, 57, 312–333.

Mehlmann, A (1988). *Applied Differential Games*. New York: Kluwer Academic Publishers.

Meijdam, AC and AJ de Zeeuw (1986). On expectations, information and dynamic game equilibria. *Journal of Economic Dynamics and Control*, 10, 63–66.

Miller, J and J Andreoni (1991). Can evolutionary dynamics explain free riding in experiments? *Economics Letters*, 36, 9–15.

Miller, M and M Salmon (1985). Dynamic games and the time inconsistency of optimal policies in open economies. *Economic Journal*, 95, 124–137.

Mino, K (2001a). Optimal taxation in dynamic economies with increasing returns. *Japan and the World Economy*, 13, 235–254.

Mino, K (2001b). On time consistency in Stackelberg differential games, Working Paper, Faculty of Economics, Kobe University.

Miravete, EJ (2003). Time-consistent protection with learning by doing. *European Economic Review*, 47, 761–790.

Miyagiwa, K and Y Ohno (1999). Credibility of protection and incentive to innovate. *International Economic Review*, 40, 143–163.

Moledina, AA, JS Coggins, S Polasky and C Costello (2003). Dynamic environmental policy with strategic firms: Prices versus quantities. *Journal of Environmental Economics and Management*, 45, 356–376.

Murdoch, JC and T Sandler (1986). Complementarity, free riding, and the military expenditures of NATO allies. *Journal of Public Economics*, 25, 83–101.

Newbery, D (1976). A Paradox in Tax Theory: Optimal Tariffs on Exhaustible Resources, Unpublished manuscript.

Newbery, D (1981). Oil prices, cartels, and the problem of dynamic inconsistency. *Economic Journal*, 91, 617–646.

Nkuiya-Mbakop, R Bruno (2009). The Effects of the Length of the Period of Commitment on the Size and the Stability of International Environmental Agreements, Typescript, Université de Montréal.

Nordhaus, WD (1994). *Managing the Global Commons: The Economics of Climate Change*. Cambridge, MA: MIT Press.

Nordhaus WD and Z Yang (1996). A regional dynamic general equilibrium model of alternative climate change strategies. *American Economic Review*, 86, 741–756.

Obstfeld, M (1989). Dynamic Seigniorage theory: An exploration, NBER Working Paper 2869.

Oksendal, B (2007). *Stochastic Differential Equations*. New York: Springer.

Olson, L (1988). Strategic considerations in invention and innovation: The case of natural resources revisited. *Econometrica*, 56, 841–849.

Ostrom, E (1990). *Governing the Commons: The Evolution of Institutions for Collective Actions*. Cambridge, UK: Cambridge University Press.

Oudiz, G and J Sachs (1984). International policy coordination in dynamic macroeconomic models. In *International Economic Policy Coordination*, W Buiter and R Marston (eds.), New York: Cambridge University Press.

Papavassilopoulos, G and S Cruz (1980). Sufficient conditions for Stackelberg and Nash strategies with memories. *Journal of Optimization Theory and Applications*, 31, 253–260.

Pearce, D and E Stacchetti (1997). Time-consistent taxation by a government with redistributive goal. *Journal of Economic Theory*, 72, 282–305.

Pethig, R (1992). *Conflicts and Cooperations in Managing Environmental Resources*. Berlin: Springer-Verlag.

Petrosjan L and G Zaccour (2003). Time-consistent Shapley value allocation of pollution cost reduction. *Journal of Economic Dynamics and Control*, 27, 381–398.

Pezzey, J (1994). Theoretical Essays on Sustainability and Environmental Policy, PhD thesis, University of Bristol.

Philippoponulos, A and G Economides (2005). Are Nash tax rates too low or too High? The role of endogenous growth. *Review of Economic Dynamics*, 6, 37–53.

Pichler, P and G Sorger (2009). Wealth distribution and aggregate time preference: Markov-perfect equilibria in a Ramsey economy. *Journal of Economic Dynamics and Control*, 33(1), 1–14.

Pitchford JD and SJ Turnovsky (1977). *Applications of Control Theory to Economic Analysis*. Amsterdam: North-Holland.

Ploeg, F van der (1986). Inefficiency of oligopolistic resource markets with isoelastic demand, zero extraction costs and stochastic renewal. *Journal of Economic Dynamics and Control*, 10, 309–314.

Ploeg, F van der (1987). Inefficiency of credible strategies in oligopolistic resource markets with uncertainty. *Journal of Economic Dynamics and Control*, 11, 123–145.

Ploeg, F van der and C Withagen (1991). Pollution control and the Ramsey problem. *Environmental and Resource Economics*, 1, 215–236.

Ploeg, F van der and AJ De Zeew (1990). Perfect equilibrium in a model of competitive arms accumulation. *International Economic Review*, 31(1), 131–146.

Ploeg, F van der (2010a). Voracious transformation of a common natural resource into productive capital. *International Economic Review*, 51(2), 365–381.

Ploeg, F van der (2010b). Rapacious resources depletion, excessive investment and insecure property rights. CESifo Working Paper No. 2981, University of Munich.

Ploeg, F van der and AJ de Zeeuw (1992). International aspects of pollution control. *Environmental and Resource Economics*, 2, 117–139.

Plourde, C and D Yeung (1993). Harvesting a transboundary replenishable fish stock: A non-cooperative game solution. *Marine Resource Economics*, 6, 57–71.

Polasky, S (1990). Exhaustible resource oligopoly: Open-loop and Markov perfect equilibria, Boston College Working Paper 199.

Polasky, S (1992). Do oil producers acts as oil igopolists? *Journal of Environmental Economics and Management*, 23, 216–247.

Pohjola, M (1983). Nash and Stackelberg solutions in a differential game model of capitalism. *Journal of Economic Dynamics and Control*, 6, 173–186.

Pohjola, M (1986). Applications of dynamic game theory to macroeconomics. In *Dynamic Games and Applications in Economics*. Başar, T (ed.), pp. 103–133. Berlin and New York: Springer.

Prescott, E (1977). Should control theory be used for economic stabilization? In *Optimal Policies, Control Theory, and Technology Exports, Carnegie-Rochester Conference Series on Public Policy*, 7, 13–38.

Quyen, NV (1988). The optimal depletion and exploration of a nonrenewable resources. *Econometrica*, 56(6), 1467–1471.

Ramsey, F (1928). A mathematical theory of saving. *Economic Journal*, 38, 543–559.

Rawls, J (1971). *A Theory of Justice*. (1st Ed.). Cambridge, MA: The Belknap Press of the Harvard University Press.

Rawls, J (1999). *A Theory of Justice* (Revised Edition). Cambridge, Massachusetts: The Belknap Press of the Harvard University Press.

Reinganum, JF (1981a). Dynamic games of innovation. *Journal of Economic Theory*, 25, 21–41.

Reinganum, JF (1981b). On the diffusion of new technology: A game theoretic approach. *Review of Economic Studies*, 48, 385–405.

Reinganum, JF (1982a). A dynamic game of R & D: Patent protection and competitive behavior. *Econometrica*, 50, 671–688.

Reinganum, JF (1982b). Strategic search theory. *International Economic Review*, 23, 1–17.

Reinganum, JF (1985). Innovation and industry evolution. *Quarterly Journal of Economics*, 100, 81–99.

Reinganum, JF (1989). The timing of innovation: Research, development, and diffusion. In *Handbook of Industrial Organization*, Vol 1, R Schmalensee and RD Willig, (eds.). Amsterdam: Elsevier.

Reinganum, JF and NL Stokey (1985). Oligopoly extraction of a common property natural resource: The importance of period of commitment in dynamic games. *International Economic Review*, 26, 161–173.

Reiss, H (1970). *Kant's Political Writings*. Cambridge, UK: Cambridge University Press.

Reynolds, S (1987). Capacity investment, preemption, and commitment in an infinite horizon model. *International Economic Review*, 28, 69–88.

Reynolds, S (1991). Dynamic oligopoly with capacity adjustment costs. *Journal of Economic Dynamics and Control*, 15, 491–514.

Rodiguez, A (1981). Rawls' maximim criterion and time consistency: A generalization. *Review of Economic Studies*, 48, 599–605.

Rogoff, K (1985). The optimal degree of commitment to an intermediate monetary target. *Quarterly Journal of Economics*, 100, 1169–1189.

Rogoff, K (1990). Equilibrium political budget cycles. *American Economic Review*, 80, 21–36.

Rogoff, K and A Sibert (1988). Elections and macroeconomic policy cycles. *Review of Economic Studies*, 55, 1–16.

Roos, CF (1925). A mathematical theory of competition. *American Journal of Mathematics*, 46, 163–175.

Roos, CF (1927). A dynamic theory of economics. *Journal of Political Economy*, 35, 632–656.

Rosen, JB (1965). Existence and uniqueness of equilibrium points for concave N-person games. *Econometrica*, 33, 520–534.

Rowat, C (2007). Non-linear strategies in a linear-quadratic differential games. *Journal of Economic Dynamics and Control*, 31, 3179–3202.

Rubio, S (2005). Tariff agreements and non-renewable resource international monopolies: Prices versus quantity, Working Paper WP-AD- 2005-10, University of Valencia.

Rubio, S and B Casino (2003). Strategic behavior and efficiency in common property extraction of ground water. *Environmental and Resource Economics*, 26, 73–87.

Rubio, S and B Casino (2005). Self-enforcing international environmental agreements with a stock pollutant. *Spanish Economic Review*, 7, 89–109.

Rubio, S and L Escriche (2001). Strategic pigouvian taxation, stock externalities and polluting non-renewable Resources. *Journal of Public Economics*, 79, 297–313.

Rubio, S and A Ulph (2007). An infinite horizon model of dynamic membership of international environmental agreements. *Journal of Environmental Economics and Management*, 54, 296–310.

Rustichini, A (1992). Second-best equilibria for games in joint exploitation of a productive asset. *Economic Theory*, 2, 191–196.

Salant, S (1976). Exhaustible resource and industrial structure: A Nash–Cournot approach to the world oil market. *Journal of Political Economy*, 84, 1079–1094.

Salant, S (1982). Imperfect competition in the international energy market a computerized Nash-Cournot model, *Operations Research*, 30, 252–280.

Salant, S, S Switzer and R Reynolds (1983). Losses from horizontal mergers: The effects of an exogenous change in industry structure on Nash-Cournot equilibrium. *Quarterly Journal of Economics*, 98(2), 185–203.

Salo, S and O Tahvonen (2001). Oligopoly equilibria in non-renewable resource markets. *Journal of Economic Dynamics and Control*, 25, 671–702.

Sandal, L and S Steinshamn (2004). Dynamic Cournot competitive harvesting of a common pool resource. *Journal of Economic Dynamics and Control*, 28, 1781–1799.

Sannikov, Y (2008). A continuous-time version of the principal-agent problem. *Review of Economic Studies*, 75, 957–984.

Sappington, D and DS Sibley (1988). Regulating without cost information: The incremental surplus subsidy scheme. *International Economic Review*, 29(2), 297–306.

Schaefer, MB (1957). Some considerations of population dynamics and economics in relation to the management of marine fisheries. *The Fisheries Research Board of Canada*, 14, 669–681.

Shubik, M and W Whitt (1973). Fiat money in an economy with one nondurable good and no credit. In *Topics in Differential Games*, A Blaquiere (ed.). Amsterdam: North Holland.

Shimomura, K (1991). The feedback equilibria of a differential game of capitalism. *Journal of Economic Dynamics and Control*, 15, 317–338.

Shimomura, K and D Xie (2008). Advances on Stackelberg open-loop and feedback strategies. *International Journal of Economic Theory*, 4, 115–133.

Shleifer, A (1985). A theory of yardstick competition. *Rand Journal of Economics*, 16, 319–327.

Sieper, E and P Swan (1973). Monopoly and competition in the market for durable goods. *Review of Economic Studies*, 40(3), 333–351.

Simaan, M and J Cruz (1973a). On the Stackelberg strategy in non-zero sum games. *Journal of Optimization Theory and Applications*, 11(5), 533–555.

Simaan, M and J Cruz (1973b). Additional aspects of Stackelberg strategy in nonzero-sum games. *Journal of Optimization Theory and Applications*, 1(6), 613–626.

Simaan, M and J Cruz (1975). Formulation of Richardson's model of arms race from a differential game viewpoint. *Review of Economic Studies*, 42, 67–77.

Simon, LK and MB Stinchcombe (1989). Extensive form games in continuous time: Pure strategies. *Econometrica*, 57, 1171–1214.

Sinn, H-W (1984). Common property resources, storage facilities, and ownership structures: A Cournot model of the oil market. *Economica*, 51(23), 235–252.

Sinn, H-W (2008). Public policies against global warming: A supply-side approach. *International Tax and Public Finance*, 15(4), 360–394.

Smith, JL (2009). World oil: Market or Mayhem? *Journal of Economic Perspectives*, 23(3), 145–164.

Solow, RM (1974). Intergenerational equity and exhaustible resources. *Review of Economic Studies, Symposium on the Economics of Exhaustible Resources*, pp. 29–46.

Sorger, G (2000). Income and wealth distribution in a simple model of growth. *Economic Theory*, 16, 23–42.

Sorger, G (2002). On the long-run distribution of capital in the Ramsey model. *Journal of Economic Theory*, 105, 226–243.

Sorger, G (2005). A dynamic common property resource problem with amenity value and extraction costs. *International Journal of Economic Theory*, 1, 3–19.

Soubeyran, A (1988). The closed-loop motive for an export tax in a dynamic Cournot game of international market rivalry. *Paper presented at the International Conference on International Policy Coordination*, Aix-en-Provence.

Spence, AM (1979). Investment strategy and growth in a new market. *Bell Journal of Economics*, 10, 1–19.

Spence, AM (1981). The learning curve and competition. *Bell Journal of Economics*, 12, 49–70.

Staiger, RW and G Tabellini (1987). Discretionary trade policy and excessive protection. *American Economic Review*, 77, 823–837.

Stachurski, J (2002). Stochastic optimal growth with unbounded stock. *Journal of Economic Theory*, 106, 40–65.

Starr, AW and YC Ho (1969a). Nonzero-sum differential games. *Journal of Optimization Theory and Applications*, 3(3), 184–206.

Starr, AW and YC Ho (1969b). Further properties of nonzero-sum differential games. *Journal of Optimization Theory and Applications*, 3(4), 207–219.

Stigler, G (1965). The tenable range of functions of local government. In *Private Wants and Public Needs*, ES Phelps (ed.), New York: Norton pp. 167–176.

Stiglitz, J (1976). Monopoly and the rate of extraction of exhaustible resources. *American Economic Review*, 66, 655–661.

Stokey, NL (1981). Rational expectations and durable good pricing. *Bell Journal of Economics*, 12, 112–128.

Suga, K (2004). On the Role of Externalities in the Arrow-Dasgupta Economy, unpublished manuscript, Waseda University.

Sugden, R (1985). Consistent conjectures with voluntary contributions to public goods. *Journal of Public Economics*, 27, 117–124.

Sundaram, RK (1989). Perfect equilibrium in non-randomized strategies in a class of symmetric dynamic games. *Journal of Economic Theory*, 47, 153–177.

Tahvonen, O (1995). International CO_2 taxation and the dynamics of fossil fuel markets. *International Tax Public Finance*, 2, 261–278.

Tahvonen, O (1996). Trade with polluting non-renewable resources. *Journal of Environmental Economics and Management*, 30, 1–17.

Takayama, T and M Simaan (1984). Differential game theory policies for consumption regulation of renewable resources. *IEEE Transactions on Systems, Man and Cybernetics*, 14, 764–766.

Tidball, M and G Zaccour (2005). An environmental game with coupling constraints. *Environmental Modeling and Assessment*, 10, 153–158.

Tirole, J (1988). *The Theory of Industrial Organization*. Cambridge, MA: MIT Press.

Tornell, A (1991). Time-inconsistency of protectionist programs. *Quarterly Journal of Economics*, 106, 963–974.

Tornell, A (1997). Economic growth and decline with endogenous property rights. *Journal of Economic Growth*, 2, 219–250.

Tornell, A and P Lane (1996). Power, growth, and the voracity effect. *Journal of Economic Growth*, 1, 213–241.

Tornell, A and P Lane (1999). The voracity effect. *American Economic Review*, 89, 22–46.

Tornell, A and A Velasco (1992). The tragedy of the commons and economic growth: Why does capital flow from poor to rich countries? *Journal of Political Economy*, 100, 1208–1231.

Tolwinski, B (1982) A concept of cooperative equilibrium for dynamic games. *Automatica*, 18, 431–447.

Tolwinski, B, A Haurie and G Leitmann (1986). Cooperative equilibria in differential games. *Journal of Mathematical Analysis and Applications*, 119, 182–202.

Tsutsui, S and K Mino (1990). Non-linear strategies in dynamic duopolistic competition with sticky prices. *Journal of Economic Theory*, 52, 136–161.

Turnovsky, SJ (1986). Optimal tariffs in consistent conjectural variations equilibrium. *Journal of International Economics*, 21, 301–312.

Turnovsky, SJ (2000). *Methods of Macroeconomic Dynamics*. Cambridge, MA: MIT Press.

Turnovsky, SJ, T Başar and V d'Orey (1988). Dynamic strategic monetary policies and coordination in interdependent economies. *American Economic Review*, 78, 341–361.

Turnovsky, SJ and V d'Orey (1986). Monetary policies in interdependent economies: A strategic approach. *Economic Studies Quarterly*, 37, 114–133.

Vogelsang, I and J Finsinger (1979). A regulatory adjustment process for optimal pricing by multiproduct monopoly firms. *Bell Journal of Economics*, 10(1), 157–171.

Warr, P (1983). The private provision of a public god is independent of the distribution of income. *Economics Letters*, 13, 207–211.

Weitzman, ML (1974a). Prices versus quantities. *Review of Economic Studies*, 41, 477–491.

Weitzman, ML (1974b). Free access vs. private ownership as alternative systems for managing common property. *Journal of Economic Theory*, 8(2), 225–234.

Weinstein, MC and R Zeckhauser (1975). The optimal consumption of depletable natural resources. *Quarterly Journal of Economics*, 89, 371–392.

Wildasin, DE (2002). Fiscal competition in space and time? *Journal of Public Economics*, 87, 2571–2588.

Wildasin, DE (2008). Fiscal Competition for Imperfectly Mobile Labor and Capital: A Comparative Dynamic Analysis, CESifo Working Paper 2808, University of Munich.

Wirl, F (1994). Pigouvian taxation of energy for flow and stock externalities and strategic, non-competitive energy pricing. *Journal of Environmental Economics and Management*, 26, 1–18.

Wirl, F (1995). The exploitation of fossil fuels under the threat of global warming and carbon taxes: A dynamic game approach. *Environmental and Resource Economics*, 5, 333–352.

Wirl, F (1996). Dynamic voluntary provision of public goods: Extension to non-linear strategies. *European Journal of Political Economy*, 12, 555–560.

Wirl, F (2007a). Energy prices and carbon taxes under uncertainty about global warming. *Environmental and Resource Economics*, 36, 313–340.

Wirl, F (2007b). Do multiple Nash equilibria in Markov strategies mitigate the tragedy of the commons? *Journal of Economic Dynamics and Control*, 31, 3723–3740.

Wirl, F (2008). Tragedy of the commons in a stochastic game of a stock externality. *Journal of Public Economic Theory*, 10(1), 99–124.

Wirl, F and E Dockner (1995). Leviathan governments and carbon taxes: Costs and potential benefits. *European Economic Review*, 39, 1215–1236.

Xepapadeas, AP (1992). Environmental policy design and dynamic nonpoint-source pollution. *Journal of Environmental Economics and Management*, 23, 22–39.

Xie, D (1997). On time consistency: A technical issue in Stackelberg differential games. *Journal of Economic Theory*, 6, 412–430.

Yang, Z (2003). Reevaluation and renegotiation of climate change coalitions: A sequential closed-loop game approach. *Journal of Economic Dynamics and Control*, 27, 1563–1594.

Yanase, A (2005). Pollution control in open economies: Implications of within-period interactions for dynamic game equilibrium. *Journal of Economics*, 84(3), 277–311.

Yanase, A (2007). Dynamic games of environmental policy in a global economy: Taxes versus quotas. *Review of International Economics*, 15(3), 592–611.

Yeung, DWK (1992). A differential game of industrial pollution management. *Annals of Operations Research*, 37, 297–311.

Yeung, DWK (2007). Dynamically consistent cooperative solution in a differential game of transboundary industrial pollution. *Journal of Optimization Theory and Applications*, 134, 143–160.

Yeung, DWK and LA Petrosyan (2008). A cooperative stochastic differential game of transboundary industrial pollution. *Automatica*, 44, 1532–1544.

Yin, X (2004). Voluntary import expansions with non-stationary demand. *Canadian Journal of Economics*, 37(4), 1084–1096.

Zagonari, F (1998). International pollution problems: Unilateral initiatives by environmental groups in one country. *Journal of Environmental Economics and Management*, 36, 46–69.

INDEX